Northern Sparks

Leonardo

Sean Cubitt, Editor-in-Chief

Erica Hruby, Editorial Director, Leonardo/ISAST

Synthetics: Aspects of Art and Technology in Australia, 1956–1975, Stephen Jones, 2011

Hybrid Cultures: Japanese Media Arts in Dialogue with the West, Yvonne Spielmann, 2012

Walking and Mapping: Artists as Cartographers, Karen O'Rourke, 2013

The Fourth Dimension and Non-Euclidean Geometry in Modern Art, revised edition, Linda Dalrymple Henderson, 2013

Illusions in Motion: Media Archaeology of the Moving Panorama and Related Spectacles, Erkki Huhtamo, 2013

Relive: Media Art Histories, edited by Sean Cubitt and Paul Thomas, 2013

Re-collection: Art, New Media, and Social Memory, Richard Rinehart and Jon Ippolito, 2014

Biopolitical Screens: Image, Power, and the Neoliberal Brain, Pasi Väliaho, 2014

The Practice of Light: A Genealogy of Visual Technologies from Prints to Pixels, Sean Cubitt, 2014

The Tone of Our Times: Sound, Sense, Economy, and Ecology, Frances Dyson, 2014

The Experience Machine: Stan VanDerBeek's Movie-Drome and Expanded Cinema, Gloria Sutton, 2014

Hanan al-Cinema: Affections for the Moving Image, Laura U. Marks, 2015

Writing and Unwriting (Media) Art History: Erkki Kurenniemi in 2048, edited by Joasia Krysa and Jussi Parikka, 2015

Control: Digitality as Cultural Logic, Seb Franklin, 2015

New Tendencies: Art at the Threshold of the Information Revolution (1961–1978), Armin Medosch, 2016

Screen Ecologies: Art, Media, and the Environment in the Asia-Pacific Region, Larissa Hjorth, Sarah Pink, Kristen Sharp, and Linda Williams, 2016

Pirate Philosophy: For a Digital Posthumanities, Gary Hall, 2016

Social Media Archeology and Poetics, edited by Judy Malloy, 2016

Practicable: From Participation to Interaction in Contemporary Art, edited by Samuel Bianchini and Erik Verhagen, 2016

Machine Art in the Twentieth Century, Andreas Broeckmann, 2016

Here/There: Telepresence, Touch, and Art at the Interface, Kris Paulsen, 2017

Voicetracks: Attuning to Voice in Media and the Arts, Norie Neumark, 2017

Ecstatic Worlds: Media, Utopias, Ecologies, Janine Marchessault, 2017

Interface as Utopia: The Media Art and Activism of Fred Forest, Michael F. Leruth, 2017

Making Sense: Cognition, Computing, Art, and Embodiment, Simon Penny, 2017

Weather as Medium: Toward a Meteorological Art, Janine Randerson, 2018

Laboratory Lifestyles: The Construction of Scientific Fictions, edited by Sandra Kaji-O'Grady, Chris L. Smith, and Russell Hughes, 2018

Invisible Colors: The Arts of the Atomic Age, Gabrielle Decamous, 2018

Virtual Menageries: Animals as Mediators in Network Cultures, Jody Berland, 2019

From Fingers to Digits: An Artificial Aesthetic, Ernest Edmonds and Margaret A. Boden, 2019

Material Witness: Media, Forensics, Evidence, Susan Schuppli, 2020

Tactics of Interfacing: Encoding Affect in Art and Technology, Ksenia Fedorova, 2020

Giving Bodies Back to Data: Image-Makers, Bricolage, and Reinvention in Magnetic Resonance Technology, Silvia Casini, 2021

A Biography of the Pixel, Alvy Ray Smith, 2021

Living Books: Experiments in the Posthumanities, Janneke Adema, 2021

Art in the Age of Learning Machines, Sofian Audry, 2021

Design in Motion: Film Experiments at the Bauhaus, Laura Frahm, 2022

Northern Sparks: Innovation, Technology Policy, and the Arts in Canada from Expo 67 to the Internet Age, Michael Century, 2022

See http://mitpress.mit.edu for a complete list of titles in this series.

Northern Sparks

Innovation, Technology Policy, and the Arts in Canada from Expo 67 to the Internet Age

Michael Century

The MIT Press
Cambridge, Massachusetts
London, England

The MIT Press would like to thank the anonymous peer reviewers who provided comments on drafts of this book. The generous work of academic experts is essential for establishing the authority and quality of our publications. We acknowledge with gratitude the contributions of these otherwise uncredited readers.

This book was set in Stone Serif and Stone Sans by Westchester Publishing Services. Printed and bound in the United States of America.

Library of Congress Cataloging-in-Publication Data

Names: Century, Michael, author.
Title: Northern sparks : innovation, technology policy, and the arts in Canada from Expo 67 to the Internet age / Michael Century.
Description: Cambridge, Massachusetts : The MIT Press, [2022] | Series: Leonardo | Includes bibliographical references and index.
Identifiers: LCCN 2021033966 | ISBN 9780262045001 (paperback)
Subjects: LCSH: Technological innovations—Canada—History—20th century. | Digital media—Canada—History—20th century. | Technology—Government policy—Canada—History—20th century. | Technology and the arts—History—20th century.
Classification: LCC T173.5. C2 C46 2022 | DDC 302.23/10971—dc23
LC record available at https://lccn.loc.gov/2021033966

10 9 8 7 6 5 4 3 2 1

For Barbara Todd

Contents

Series Foreword

Leonardo/The International Society for the Arts, Sciences and Technology fosters transformation at the nexus of art, science, and technology because complex problems require creative solutions. The Leonardo book series shares these aims of artistic and scientific experimentation, publishing books to define problems and discover solutions, to critique old knowledge and create the new.

In the early twentieth century, the arts and sciences seemed to interact instinctively. Modern art and modern poetry were automatically associated with relativity and quantum physics, as if the two were expressions of a single zeitgeist. At the end of the Second World War, once again it seemed perfectly clear that avant-garde artists, architects, and social planners would join cyberneticists and information theorists to address the problems of the new world order and to create new ways of depicting and understanding its complexity through shared experiences of elegance and experiment. Throughout the twentieth century, the modern constantly mixed art and science.

In the twenty-first century, though, we are no longer modern but contemporary, and now the wedge between art and science that C. P. Snow saw emerging in the 1950s has turned into a culture war. Governments prefer science to arts education, yet stand accused of ignoring or manipulating science. The arts struggle to justify themselves in terms of economic or communicative efficiency that devalues their highest aspirations. And yet never before have artists, scientists, and technologists worked together so closely to create individual and collective works of cultural power and intellectual grace. *Leonardo* looks beyond predicting dangers and challenges, beyond even planning for the unpredictable. The series publishes books that are both timely and of enduring value—books that address the perils of our time, while also exploring new forms of beauty and understanding.

Sean Cubitt, Editor-in-Chief, Leonardo Book Series

Erica Hruby, Editorial Director, Leonardo/ISAST

Preface

As an American growing up in western Canada during the 1960s, in the shadow of the oil wells that brought my family there in the first place, my prevailing orientation was to the south. I was drawn more to the revolutionary allures of America, awful yet compelling in those turbulent years. When my liberal parents chose to keep us in Canada during the height of the Vietnam War, I initially felt sidelined, remote from the heart of the action. Thus it was with the fervor of a convert that I later adopted a way of identifying as a Canadian in relational terms, ever comparing, never fixed, with a dual sense of proximity and distance born out of my own initial ambivalence.

In a similar way, this book reveals how a uniquely Canadian story can provide a fresh angle on a well-established narrative. *Northern Sparks* is about the intersecting histories of technological innovation, new media arts, and government policies in Canada from the 1960s to the early 1990s. It portrays this period as an "episode of light" sparked by an intense national awakening following the international success of the media-rich 1967 world's fair in Montréal during the nation's centennial year. The opening conjuncture of the arts, technology, and political consciousness elicited an exceptionally broad range of outcomes that go beyond notable artworks or new technological systems and tools. Poised as a "counter-environment" to the great powers, in Marshall McLuhan's phrase, Canada's unique experience of the transitional decades into the information age was grounded in a technological ethos that emphasized sensorial immediacy, embodied interaction, and improvisatory expression. This alternative ethos was situated between a pair of distinct yet inextricably bound forces, one national-political and proper to Canada, the other techno-mediatic and global in scale. The unraveling of these forces by the late millennium reveals innovation itself as a complexly drawn process comprised of multiple layers with fluctuating degrees of synchronization.

I am grateful to the many individuals who generously offered their insights and encouragement over the course of this book's long gestation. Justin Sully provided the indispensable perspective as well as editorial acumen needed to complete it. For the conversations that helped launch the project, I thank Thierry Bardini, Bill Buxton, Robin Mansell, David Neice, B. W. Powe, Catherine Richards, Joan Shigekawa, Ed Steinmueller, and Will Straw. McGill University's Centre for Research on Canadian Cultural Industries and Institutions gave me the institutional home that incubated the project, Hexagram at Concordia University supported a research residency, and the School of Humanities, Arts and Social Sciences at Rensselaer Polytechnic Institute provided a conducive interdisciplinary milieu in which to combine my interests in new media history and theory with policy studies. Nancy Campbell, Tom Sherman, Ron Graham, Barry Truax, Robert Nideffer, Zach Layton, and Ezra Teboul offered incisive commentary along the way. I am particularly thankful to the individuals who gave their time for interviews; the full list is provided in Appendix A.

1 An Episode of Light: Canada from 1967 to 1992

This book offers an account of the conjuncture of media technology, innovation policy, and creative art practice that emerged and flowered in Canada from the late 1960s to the early 1990s. Aligning roughly with the chronology of the installation and first mass adoption of digital media and computation in Canada, this is a story of glittering innovations in new media, as well as squandered opportunities and unrealized futures that came into view during these years. Taken together, the cases around which the book is organized highlight the decisive, catalytic role played by artists in what appears today as a sustained *episode of light*,[1] a period of intense creativity in the arts and media technology in Canada that can be dated from 1967 to about 1992. Out of this conjuncture also emerged new art forms, interdisciplinary configurations, prescient speculative designs, and novel institutional and policy frameworks, all of which were marked by a remarkably sustained and reflexive deliberation on the relationships between humans and their extensions through media.[2] This was a period in which media innovation became tightly enmeshed in a critical reimagining of a uniquely Canadian nationalism, during which Canada achieved a kind of distinction internationally, not only in what is acknowledged to be an influential Canadian "discourse on culture and technology,"[3] but in a set of creative practices in both domains that emphasized sensorial immediacy, embodied interaction, and improvisational expression. These values comprised a shared culture, or what I will term an "alternative technological ethos": a common way of apprehending technology as a malleable material, open to revision by artists in use, and subject to both conceptual and functional reimaginings.[4] It is toward a clearer understanding of this alternative technological ethos—its historical and political-economic contexts, its defining qualities and practices, its achievements and failures, and the lessons it might offer—that this book aspires.

This "episode of light" begins with the heady, even ecstatic enthusiasms of the World Exposition at Montréal, which marked the nation's centenary in 1967. It spans the following quarter-century, as new technologies for computing and networking begin to seep slowly into laboratories and studios, spreading with increasing rapidity across the country. The period it describes corresponds closely with the build-up or installation phase of what will here be called the "Information Paradigm"—a shorthand to indicate not just the technological underpinnings of computers, software, and networks, but also the artistic expressions that worked within or challenged the limitations of these developing infrastructures as cultural terrains.

Canada had a well-established set of institutions and technocultural discourses already in place as the installation phase of the Information Paradigm was getting started. Public policies supported cultural and scientific institutions, such as the National Research Council and the National Film Board, that were open to interdisciplinary experimentation. Within the federal government, the connections between cultural and communications policy were recognized, if not coherently worked out, when they were placed under a single government department for a brief fifteen-year period. New university-based interdisciplinary academic programs (like Simon Fraser University's Faculty of Interdisciplinary Studies), departments receptive to artistic research directions (like the Computer Science Department at the University of Toronto), and art schools with both traditional and experimental wings (like the Ontario College of Art and the Banff Centre in Alberta) were able to harbor, incubate, and endorse new approaches, often also assuming the role of official advocate, articulating a coherent rationale for recognizing artists as valued actors in techno-aesthetic innovation. Independent, artist-led initiatives were also fostered, enabling the formation of a publicly funded national network of alternative centers, many of them organized around cooperative exploration of both older communication media and the emerging digital technologies. These preexisting policy and institutional frameworks were themselves part of a long and widely remarked-upon history of the coevolution of communication and nationalism in Canada.[5] This intersecting history of media and national consciousness also contributes to the intensity of the book's starting point in 1967, when the nation was experiencing a palpable political reinvention alongside its centennial celebration, and artists were taking on the slippery challenge of defining what it might mean to speak of a distinctly Canadian culture.

In the 1970s and 1980s, the installation phase of the Information Paradigm began to accelerate and its potential threats and opportunities came more sharply into view. Against this backdrop, there arose among some Canadian artists a newfound clarity in their efforts to analyze and critically intervene in what were already beginning to be recognized as potentially restrictive standardizations, as corporate consolidation over the information interfaces and applications spread during the last decades of the century. Scattered throughout this book are cases that address some of these critical interventions, in which artists issued challenges to and proposed solutions for a wide range of problems: the design of humane soundscapes, the "transparency" of user interfaces, the configuration of creative practices in application software, the nature of global interactive networks, and the place of the human body in the emerging field of virtual reality, to name a few. These interventions were in part fueled intellectually by the homegrown Canadian discourses on technology and culture, typified but by no means limited to Marshall McLuhan's ideas, which privileged artistic insights as a creative and critical resource for understanding the deeper and profoundly ambivalent long-term effects of new media.

With the advent of the personal computer in the 1980s and the World Wide Web in the early 1990s, what I will call the installation phase of the Information Paradigm comes to an end and its mass deployment phase begins in earnest. It is at this historical juncture—the dawn of the internet age, around 1992—that this book comes to a close. The deployment phase of this cluster of enabling technologies coincided, paradoxically, with the cooling of Canada's burst of self-confidence as a modern nation, when a host of existential crises within its own borders, and the rampant pressures of a neoliberal international order beyond them, ushered in a new era.

At the political level, Canada in the later 1980s was consumed by the challenges to national unity occasioned by a rising Québec independence movement, and by the perceived threats to economic autonomy and national treatment of cultural funding arising out of the negotiations for free trade with the United States after 1984. The country descended from an apogee moment of exuberance and national self-discovery in 1967 to existential crisis and doubts about its very viability as a nation. This declining domestic arc occurred alongside the hastening *rise* in adoption of digital technologies at large, building by the early 1990s toward the transnational corporate-led surge into digital media convergence and deregulation. The resonant energy unleashed through the intersection of Canadian cultural self-refashioning and experimental media practices was attenuated just as the information deluge of the 1990s swept over the planet, bringing to Canada (as everywhere else) a corporatized version of digital culture. It was within the maelstrom of this "globalized" techno-environment that digital media innovation itself became a deliberate object of research policy in Canada, though a confused one with frayed connections to the kinds of open-ended, improvisatory cultural practices that had come before. This disjuncture at the official level between research and cultural policy occurred while rising consciousness of indigenous histories and cultural diversities was further undermining any lingering credence in a unitary pan-Canadian identity defined in terms of advanced media technologies.[6]

In summary, the flowering after 1967 of a media-inflected national consciousness in Canada coincided with the transitional years into the Information Paradigm. *Northern Sparks* contributes to national studies of innovation systems by offering a detailed historical account of the epochal transition to digital culture in Canada, emphasizing the significance of the creative arts to technological development, as well as to broader conceptual understandings of new media innovation at large. Set against our present moment, rife with denunciations of unbridled innovation in the information industries, the consideration here of the experimental arts as both a beneficiary *and* driver of technological invention takes a different tack. Conceiving artistic creation as endogenous, rather than external, to the sources of technological innovation can provide a more complex and nuanced alternative to dominant, narrowly market-based definitions of innovation, on the one hand, and simplistic calls for the vaguely imagined "integration" of arts and technoscience, on the other. Rather, in the history offered here, arts-animated innovation in media technology is grounded in specific material practices centered around human expressiveness and sensorial immediacy that were by turns nurtured, focused, and critically reoriented across a variety of sites for collaborative knowing and making.

Within the confluence of the twinned historical forces that shape this history—a national-political reinvention and a shifting technological paradigm—Canada's alternative technological ethos thrived across the spectrum of the creative arts and sparked *multifaceted* innovation outcomes: artistic, technological, scientific, and conceptual. This "alternative" attitude toward technology is distinctive in the sense that it allows for multiple ways of making and knowing, judging and valuing to be in play at once. Innovation, as David Stark has argued, emerges from the sustained influence of multiple and even potentially "dissonant" alternatives. Considered in this way, innovation is the creative practice of recognizing and bringing to fruition the friction that results as these alternatives clash or converge.[7] Understanding a technology as a malleable material for artistic exploration is valuable particularly when the technology is itself in an indeterminate state, not yet stabilized or even recognizable as a category. The improvisatory immediacy, direct interaction, and sensorial richness that characterize the techno-creative practices of this period in Canada are relevant beyond their context precisely because of how they can help us discover, through exploratory use, the possible forms of an indeterminate technology or media environment. Insofar as the affordances of a new tool, technology, or medium are not known in advance, they remain untethered from any particular function, end, or utility. But, after a new tool or system or interface is instantiated, made concrete in a first creative instance, it can be quickly normalized, assuming an assignable instrumental purpose that then appears inevitable and essential to the technology itself.

The central claim of this book is that the alternative technological ethos in Canada was distinctive in the variety of ways it was able to shape and value new media technologies through experimental artistic uses. This variety generated innovative outcomes that went beyond the actual artworks produced, or even the stabilization of aims or requirements into valuable tools or systems that often grew from the creative process. The innovations were also conceptual and philosophical—alternative ways to think about and approach interactive computing, sound environments, networked experience, and immersive environments.

These explorations and discoveries occurred across the full spectrum of the arts. Responding to this multiplicity of forms, the book's narrative structure is organized as a systematic, though not always neatly chronological, examination of the variegated ways in which this alternative technological ethos was germinal in each medium or art form. Imagined in terms of sensorial modalities, the story proceeds from moving images to graphical and kinesthetic interaction, then to sound and music, networks, and immersive environments. The alternative technological ethos played out at a variety of scales and organizational forms, from independent artists working in studio labs to formally mandated projects in publicly funded research and cultural institutions. It begins with the methods and influence of a singular artist, Norman McLaren, who, working alone or in small teams within the National Film Board, devised a host of original film animation techniques for acclaimed experimental works that were also subsequently legible across several fields as "proto-computational" (chapter 2). In chapter 3, two separate professional and institutional perspectives come together around a single short-term artistic production by the NFB and National Research Council, which generated artistic and technological outcomes that could, in turn, be exploited in different social and economic worlds: animation production, human-computer interaction, software engineering. Chapter 4, on gesture and interactivity, rediscovers the alternative technological ethos in small teams working both within major research labs and in small studio labs. Across the five examples highlighted in this chapter, the computer itself began to be considered as a more or less capable agent. This profoundly novel recognition itself generated the common challenge of finding ways to preserve sensorial immediacy across the threshold of human-computer interaction, both as creative practice and interdisciplinary science. Chapter 5 shows how the alternative technological ethos played out in different ways within the sonic realm, profiling the performers and composers who expanded music practice through improvisatory interactivity and intermediality, and brought to life new musically inspired interdisciplinary fields, like acoustic ecology and sound art. In chapter 6, this alternative ethos is shown in a comparison of two contrasting examples of creative practitioners discovering new uses of interactive networking, the first in a series of projects initiated by independent artists, and the second involving Telidon, an ill-fated government-initiated videotext system. The final case explores how virtual reality was contested as a new artistic medium and focuses on the attempt at the Banff Centre to discover new uses for, and formulate critical analyses of, immersive visual and sonic environments.

Another main claim of this book is that the alternative ethos was resonant with and therefore amplified by its being situated within the pair of intersecting historical arcs that etch out the background of the period: the national-political and the technological-mediatic. As any formative technological era transitions toward maturity, the openness to alternative modes of making and valuing is reduced.[8] With the mass adoption of standardized information interfaces, protocols, and packaged software applications accelerating rapidly from around the mid-1980s, a definite technocultural closure was occurring just as, at the political level of Canadian arts and communication policy formation, the wider social impacts of these various kinds of standardization were becoming more difficult to discern, understand, and develop appropriate policies to address.[9] Not surprisingly, by the mid-1990s, Canada's capacity to come up with policies integrating arts, technology, and innovation actually diminished just when its unique experimental histories in this arena might have helped provide a homegrown alternative to the apparent inevitability of transnational corporate media convergence and the commodification of digital media during the dot-com boom years.

From Apogee to Crisis: Canadian Arts and Technology Policy

> The period from the mid-sixties to the mid-eighties will be seen by Canadian historians as the period of one of the most radical reinventions in the history of a country, from Expo 67 to the Charter of Rights and Freedoms. We reinvented ourselves as a multicultural bilingual nation. We had to go through a near-death experience, but we did reinvent ourselves.[10]

Michael Ignatieff, a Canadian public intellectual and erstwhile Liberal politician, is hardly alone in thus associating the beginning of this radical reinvention with the moment of its symbolic apogee: the world's fair at Montréal celebrating "Man and his World." The fair welcomed the world to Canada with verve and sophistication, most famously setting off a powerful affective charge to the throngs of attendees through its extensive use of immersive audiovisual and multimedia environments.[11] For literary scholar Donald Theall, writing at the time of Expo 67, world expositions always produce symbols of the future, and this symbol was of the "total environment" in which the juxtaposition of information from many sources—like the fair's many pavilions with multiscreen displays—elicited a visceral awareness of a "new sense of space." This new kind of media environment not only juxtaposed myriad "particles of experience" in satisfying ways, but assisted visitors to gain a wider "awareness of society as a whole," one that was "deeper and more comprehensive."[12] In short, the modern reinvention of Canada was catalyzed by a global event that ushered in the technological imaginary of the information age.

More recently, historians, film scholars, and exhibition curators have drawn out a more mixed set of messages about the exposition that go beyond its architectural landmarks and the multisensorial allures of its plethora of media displays. Architectural historian Indirbir Singh Riar points out that while Expo 67 played a crucial role in shaping Canada's "collective sense of cultural representation and global belonging," it also contained glimpses of doubt about the plausibility of any simple "abiding faith in techno-scientific salvation" in the "aftermath of global war and the terror of nuclear holocaust."[13] This ambivalence is spotlighted by Riar in several key pavilions that get less attention than the most famous ones, such as Buckminster Fuller's geodesic-domed US pavilion, or the various pavilions hosting multiscreen movies. A pavilion dedicated to First Nations history and culture was widely acknowledged for presenting a hard-hitting denunciation, for the first time at a national scale, of the suffering and exploitation of Aboriginal inhabitants of Canada.[14] Another critical perspective is seen in the pavilion "Man the Producer," which asked visitors whether they thought technology offers you "material riches or faceless conformity," or whether technology permits you "to listen to the universe or prevents you from hearing yourself."[15] Such nuance has its roots in the relatively wide intellectual input sought by Expo 67 planners from a range of Canadian scholars and artists, and is closer in fact to the actual positions held by McLuhan on the mixed effects of new media than is widely supposed.

Commenting from Expo 67 during the first five-continent satellite TV broadcast, *Our World*, McLuhan summarized how he understood the artist as a sensory repository of insight into rapid social and technological change.

> People have always been terrified of the present. The only people with enough gumption or nerve to look right at what's happening under their noses are artists. They're specialists in sensory life. They just deliberately look at the present as if they dared it to ruin them, like Perseus and the Gorgon. The artist looks into the mirror of art and says, the heck with the Gorgon's image, I'm not terrified. Most people expect, when they look at the present, to be turned into stone, as by the Gorgon. They are terrified, and therefore they prefer the rear-view mirror. Everyone who looks ahead, as it were, is in effect looking in the rear-view mirror.[16]

Just as some historians look now at Expo 67 with a more nuanced view of its many marvels, so too has the iconic year 1967 received more measured reassessments. Doug Saunders, writing in 2017 for the *Globe and Mail*, argues that 1967 is indeed "the hinge upon which modern Canadian history turns and, in certain respects, the key to understanding the challenges of the next half-century." But for Saunders, the year is better seen as a marking point when a number of sea-changes that had already been under way for decades surfaced officially, on matters ranging from women's and gay rights to aboriginal recognition and ethnic and cultural identity:

> Canada was not remade by the decisions of 1967; it was reflected by them, for the first time. What began in 1967 was official Canada beginning to catch up with the real Canada. And that is also the lesson to be carried forward to 2017: Canadians tend to be ahead of their institutions, and every few decades it is time for a dramatic catching up, like the explosion of adjustment we saw in '67.[17]

Much of the cultural infrastructure that was fundamental for the developments covered in this book had already been set in place during the several decades before Canada's breakout year. Canada's two flagship federal cultural agencies had been established before World War II and remained fundamentally unaltered under the urbane, intellectually sophisticated Prime Minister Pierre Elliott Trudeau, who led the country between 1968 and 1984, except for a short interregnum in 1979–1980. The National Film Board (NFB), established with a mandate to interpret Canada and Canadians to the world, and the Canadian Broadcasting Corporation, tasked with ensuring Canadian voices would be heard against the onslaught of US commercial radio, were both humming with innovation decades before the Trudeau years.

A 1949 blue-ribbon study, the Massey-Lévesque report, had provided a blueprint for developing arts, humanities, and social science research funding, with many recommendations for government actions and policies, including the founding of the Canada Council (effected in 1957). The Massey-Lévesque report, known officially as the Report of the Royal Commission on National Development in the Arts, Letters and Sciences, should be read, according to cultural theorist Jody Berland, as opposing the specific threats of US domination over Canada in areas like broadcasting, publishing, and scholarship, but at the same time as welcoming "internationalist" modernism in the fine arts.[18] As much as this preferential treatment of "high art" over "mass culture" infuriated some commentators (including McLuhan), the effect of the

Massey-Lévesque report was marked and comprehensive, as summarized by Tom Henighan:

> Perhaps the most amazing thing about the Massey-Lévesque Commission is that its recommendations, resulting in government actions and policies of many kinds, appear to have done the job. During the sixties, seventies, eighties, Canadian culture came to life. There were many reasons for this flowering—economic prosperity, increasing urbanization, and population diversity among them—but the establishment of the Canada Council, the commitment of government funding to the arts, was probably the single most important factor in effecting this change and in sustaining it. The growth of Canadian arts and culture during this period was amazing; even more amazing, perhaps, was the Canadian public's acceptance of this growth.[19]

Under Trudeau's first government (1968–1972), employment programs supported "Local Initiatives" driven by grassroots community needs, and "Opportunities for Youth" designated formally under cabinet mandates to promote "participation," "culture," and "engagement."[20] The bureaucrat recruited by Trudeau to direct these programs, Bernard Ostry, summarized the rationale as one of developing and strengthening "a sense of Canadian citizenship, chiefly through programs that would aid participation and assuage feelings of social injustice."[21] Before being shut down under charges of scandal, subversion, and mismanagement, these jobs-related programs contributed the initial funding for what would become the national network of artist-directed centers, a vital component of Canada's media innovation culture in the 1980s and beyond.[22]

In 1969, Trudeau also established a new federal Department of Communications with a broad mandate that would only grow over the next decade. Initially dealing mainly with the science and engineering aspects of the field, by the late 1970s the Department of Communications had expanded its scope to include portfolios supporting arts and culture, which until then had been located in the Department of the Secretary of State. This attempt at merging culture and communications, content and carrier, or software and hardware, did not endure after 1993, when the Conservative government of Kim Campbell reverted to a reactionary view of the cultural portfolios as the repositories and custodians of national heritage and identity, while the technological components were shuffled to a revamped Industry ministry incorporating science funding and policy as well.

In its initial years, however, the Department of Communications spearheaded another national study, known as the Telecommission, that came out with its own report under the title *Instant World (Sans Frontières)*. Though lacking the immediate effect of Massey-Lévesque, this report had a farsighted understanding of what would become known as "media convergence" two decades later:

> Think of a system incorporating computing, publishing, newspaper, broadcasting and library, telephone and postal services of the country, together with large slices of teaching, of operations and of many professional activities. All these each growing in its own right and subsumed in one system will outstrip in magnitude and importance any industry in which human beings have been previously engaged.[23]

Instant World also took the temperature of the intellectual and artistic community in Canada, including a chapter on the effects of new media technology on the arts and letters. This chapter, on the "Soul in the System," was the result of coast-to-coast hearings and reflected a solid dose of skepticism on the part of the artists and intellectuals consulted. Jacques Godbout, a film director from Québec, summed up this skepticism acerbically when he remarked: "The intoxication that takes hold of managers is in fact due to the fabulous quantity of knowledge which they would like to concentrate, conserve, and disseminate rationally so as to ensure that it will perpetuate the economic-technological dictatorship that provides them with a living."[24]

The report compared Godbout and others to cultural Luddites, and the tone of the chapter overall reflects something of the dichotomous view of these matters propounded by C. P. Snow in the much discussed "two cultures" debates after 1959.[25] Notwithstanding this tendency to pigeonhole humanists as "anti-technological," *Instant World* was forward-thinking on another score. The "information superhighways" envisioned, *avant la lettre*, would not just be used for broadcasting messages to the population at large, but would enable individuals to express themselves in new—albeit largely unspecified—ways. For McGill University communication scholar Marc Raboy, the report was at its most prescient when it proposed for the first time the concept of communication itself as a "human right." Anticipating a world of bidirectional communication flows, and of "user-generated content" (to use a later jargon), *Instant World* offers a tantalizing glimpse into the mindset of Trudeau-era cultural technocrats forecasting what it might mean for public policy to support such an activated, communications-rich media culture.[26]

The report's impact was muted. According to Pierre Juneau, one of the cultural mandarins recruited by Trudeau to modernize Canadian arts and communication policy,[27] the *Instant World* report helped start a federal-provincial battle over control of communications. Juneau, in an interview with the author, remarks that the Telecommission had not really dealt with the fundamental problem of that time; namely, how to ensure that there would be as much public investment in content as in hardware. "It was the start of the hardware mania," Juneau recalls, and this insight led him to convince Trudeau to put the Department of Communications and the Secretary of State together—to bridge hardware (technology, and its regulation) with "software" (cultural content) in one government department.[28]

By the time the Department of Communications had fully integrated the cultural and technology portfolios in 1979, Trudeau was briefly out of office, and pressure was building toward the first of two referendums on Québec independence (held in 1980 and 1995). The existential crisis over national unity, referred to by Ignatieff above, would become the first of two great battles won by Trudeau in the latter years of his political career. This first battle, about whether to keep Québec in Canada at all, elicited a tremendous upsurge of cultural energy as well as political activism in that province during the 1980s, which can only be touched on in this book (see chapter 5 for musical aspects arising in the context of the separatist movement). The second of Trudeau's battles, over the repatriation of Canada's Constitution from Great Britain in 1982, established a national Charter of Rights and Freedoms. It is worth recalling that Trudeau, a constitutional law professor before he entered politics in 1965, was a charismatic figure who operated at a mythic level insofar as he was able to serve as a lightning rod for diverse energies. In political terms, this was typically expressed as a vision of Canada as an open society, an original experiment that was neither American, British, nor European.[29] This openness to the world, however, was itself controversial, since the funding of Canadian "content" as such was becoming an ever more complex matter as the consideration of culture became increasingly approached as a matter of industrial policy.

A signal of the turn toward a new discourse of the "cultural industries" (later elaborated in Blair's New Labour government in Britain as "creative industries")[30] can be located in yet another large-scale Canadian effort at national consultation in the arts, this time led by Louis Applebaum and Jacques Hébert. The Report of the Federal Cultural Policy Review Committee, overseen by Applebaum and Hébert and published in 1982, showed acute awareness of the need to bolster the content *industries* (print, television, film, sound recording) against primarily American imports, while at the same time fostering "excellence" in the fine arts as such. What is most salient in this context is the report's decidedly constricted view of the relationship of art and technology. Arguing that the NFB mission should change to take on a more limited responsibility for "research and development" in the "art and science" of film production, the Applebaum-Hébert Report *also* briefly discussed the idea that the Department of Communications itself should have its own "art and technology" program. The objectives of such a program would have been the "promotion and funding of research and experimentation, and the provision of access to research results and the hardware itself to artists across the country."[31] Though this idea was not even included in the Report's actual recommendations, the Department nonetheless began to support some aspects of computerization within arts organizations, though not specifically formulated as creative research.[32] Partly in response to this rising interest inside the federal department to which it formally reported, and partly in response to what its founding director called a rising international "technological arms race" for preeminence in digital media, the Canada Council would in turn shortly start its own funding program for Media Arts, including "computer integrated media"—among the first of its kind.[33]

Canada soon faced radically increased continentalist economic pressures following Trudeau's retirement and the 1984 election of a Conservative government. National unity remained an open sore, since Trudeau's constitutional legerdemain had failed to accommodate Québec's aspirations, and the two growing tensions—over national unity, and over free trade with the United States—took much of the nation's attention from the mid-1980s onward. Much of the cultural community was preoccupied with a nationalist agenda to preserve the support infrastructures that had been built up in the half-century since the CBC was established in the height of the Great Depression. Yet another policy paper from the Department of Communications in 1987 recognized the rising economic importance of software as an industrial component and the need for Canada to develop its computer and information industries for purposes of national competitiveness.[34] But, according to University of Montréal communication scholar James Taylor, nowhere in this report—issued from the same Department of Communications whose founding came out of an attempt to merge the concerns of content and technology—was there a clear recognition of the vital importance of research on the interaction of technology and users themselves:

> The [Department] has traditionally straddled the domains of economic policy (through research, telecommunications, new technology development) and of social policy (through broadcasting and culture). This division—so reminiscent of Snow's two cultures—has tended to support a stereotype: that technology is profitable, but only secondarily social (in that one must do a "needs" analysis, and something about "impact"), while culture is social but unprofitable. What now confronts the Department is the realization that technology without sociability is not profitable: it doesn't even work. To judge from the paper that has been issued, it would seem that the problems of people still come out as an inconvenient side-effect of technological change, rather than the only way to benefit from it.[35]

By 1993, the Department of Communications was dissolved, shunting culture into the pigeonhole of national heritage and technology into that of industry and science. This end of the "under one roof" attempt at policy coordination should not be considered a failure outright, but rather as an unrealized opportunity. Overall, this is hardly surprising, given the complexity of mounting challenges of the period. An explicitly formulated "innovation policy" integrating the creative arts with scientific research and technological development was never articulated, though the intellectual ingredients that could have shaped one can clearly be discerned from the present historical vantage.

Media Arts in Innovation Policy

In order to associate the creative arts with "innovation," it is first necessary to pry the term away from its commonplace use as a business nostrum—a remedy to cure all ills. Walter Isaacson's best-selling 2014 history of the digital revolution, *The Innovators*, argues that its next phase can come only in a "true fusion" between technology and the "creative industries."[36] Educators seek to make science, technology, engineering, and mathematics more attractive by adding in the "arts," like some special sauce, to broaden technical education from "STEM to STEAM." The US National Academy of Sciences published a 2003 report, which I contributed to, urging the technology community to go "beyond productivity" by considering the role of the arts and design—here termed "creative practices"—as foundational inputs to the tech industry and university research, rather than just decorative afterthoughts.[37] In the general spirit of these well-intentioned, but not very specific, prescriptions, this book advances a historically grounded case for repositioning artistic experimentation within innovation policy studies. This section first reassesses the concept of sweeping epochal, or paradigmatic, waves of innovation in terms of the multitudinous temporalities of influence at work in the creative arts. Next it considers how the Canadian case studies of this book, which document an alternative technological ethos for research-based creative practices, may be viewed in relation to the concept of national innovation systems.

In the innovation policy literature,[38] the so-called "technological revolution" clustered around computing, software, and networks is but one in a recurrent pattern of innovation waves dating back to the first industrial revolution in England during the eighteenth century. Austrian-American economist Joseph Schumpeter developed the concept of "long waves" as recurrent patterns of industrial innovation that take the shape of a logistic curve of buildup (installation), widespread diffusion (deployment), followed by saturation and replacement as another breakthrough of invention gives rise to a new long wave.[39] The cases examined in this book took place during the buildup or installation phase of what Manuel Castells identifies as the fifth of these long waves, centered in new information and telecommunication technologies.[40] This installation period began with the earliest computers in the 1950s and 1960s, moving successively in the 1970s and 1980s toward exponential miniaturization, faster processing speeds, more user-friendly applications and interfaces, and a broader palette of audiovisual capability.

In an influential study of what was then called "Mode-2" or transdisciplinary research, published in 1994, Michael Gibbons and a group of coauthors considered the arts and humanities as outsiders to the production of new technoscientific knowledge, "standing aside as quizzical commentators who offer doom-laden prophecies or playful critiques, and as performers provide pastiche entertainment or heritage culture as a diversion from threatening complexity and volatility."[41] Gibbons also observes the way artists are implicated in the culture industry as image makers, but makes no reference to the artist as an innovative actor per se. As further observed by neo-Schumpeterian economist Christopher Freeman, artistic and cultural components of techno-economic paradigm changes have "a logic and time of their own. They may anticipate the future or recall a nostalgic vision of the past." In Freeman's history of industrial revolutions, new technologies are seen broadly to act on the "tides of cultural production," but the arts are not considered to play even an indirect role in shaping new technologies.[42] While it is true that in earlier periods, experimental artists actively took up photography, cinema, and television only after they had been established industrially, the same is not the case in digital culture. This book presents a host of cases of early adoption and co-invention by artists of digital technologies as such, but also of the creation of novel *kinds* of objects: fresh interdisciplinary configurations, like acoustic ecology and human-computer interaction; new creative genres, like soundscape composition or interactive installation art; and approaches to tool building by artists, in fields ranging from animation to music composition and interaction design.[43] A more concretely grounded and complex picture emerges, therefore, than that of futuristic anticipation or nostalgic recollection. In innovation policy terms, it will be important to track and understand transitions and overlaps, retrievals and repurposings, between different historical periods and artistic traditions as much as the ruptures and breaks.[44]

The alternative technological ethos documented in this book favored the sensorial immediacy of artistic practice as a method of exploring the emerging properties of electronic and digital technologies. The interplay of improvisatory or performative practices in film, music, or dance, on the one hand, and the formal or procedural logics of programmable systems, on the other, was generative in advancing original, human-centered understandings of the rising technologies of the fifth innovation wave. This interplay between tacit and explicit forms of knowledge played out differently in the various arts with their separate histories; musical composition and performance, for instance, have a set of notational practices enabling formalization, which differs from other performative practices like dance, and from narrative or pictorial art forms. The details of these differential affordances for leveraging the formal and procedural logics of computation are provided in the chapters to come; for present purposes it is well to note that the differentiation between the arts with respect to improvisatory immediacy and discrete formalization is akin to the concept of diverse "epistemic cultures" which emerged from laboratory studies in the sociology and anthropology of science.[45] Karin Knorr-Cetina's concept of an epistemic culture emerged from detailed analysis of work practices, and in particular the multiplicity and contingency of *inscriptions* that are used to record, stabilize, and communicate new scientific knowledge.

In a 1999 essay collection, *Picturing Science, Producing Art*, Caroline Jones and Peter Galison, argued for moving beyond the "focus on 'art' and 'science' as discrete *products*," to look at "commonalties in the practices that produce them."[46] In their subsequent efforts to realize this agenda, Galison's and Jones's work has sought to establish commonalities between twentieth-century visual art studios and experimental physics labs, emphasizing the similarities in architectural and spatial properties as well as the material and symbolic practices of the artists and scientists inhabiting them.[47] Focusing similarly on the links between institutional conditions and the specific practices afforded within them, my own research has tracked the increasingly dense formation of *hybrid* studio laboratories over the twentieth century, a tendency which accelerated during the several decades that saw the incorporation of the computer into arts practices, and into cultural applications more broadly.[48]

Canada lacked the corporate research behemoths of the United States, like ATT's Bell Labs or IBM, or the geographic concentrations of academic-industrial-military synergy like Boston's Route 128 or Silicon Valley. But it did have its own well-established national cultural and institutional infrastructure, parts of which were well-disposed to creative research in communications media, and a spate of new academic institutions with interdisciplinary missions established in the post-Baby Boom era. During the same period, France showed a marked centralized *dirigiste* approach, as has been well documented by Georgina Born in her detailed ethnography of the "rationalization" of contemporary music at IRCAM.[49] The UK, by contrast, tended primarily to locate the centers for early computer arts experiment in the polytechnic institutes, rather than in the universities, elite art schools, or music academies, reflecting a more rigid social structure and inflexible institutional ecology than Canada enjoyed.[50] Canadian cultural institutions and policies may have benefited from developing within a nation that was, to adopt a phrase from McLuhan, the "borderline case," benefiting from a close proximity to the United States and from its former imperial ties to England and France in a way that facilitated extensive knowledge flows of people, ideas, and technologies, while maintaining a certain critical distance and ambivalence about these influences. In this view, Canadian culture turned its marginality to productive use as a creative resource.[51]

In comparative studies of national innovation systems, interactive learning processes, which emphasize knowing by doing, have been put forward as alternatives to a dominant linear model of innovation that specifies a stepped sequence from science to technology to development to innovation diffusion.[52] This book does tell a specifically Canadian story of how arts-focused media innovation was fostered within a unique national cultural and research infrastructure, and it also gives rich empirical evidence of how interactive learning between different actors—artists, designers, users, and researchers—can also lead to multidimensional, not just singly acknowledged outcomes. One of these dimensions is concerned with ethical and conceptual aspects of an emerging technological *imaginary* while it is still in a formative, preparadigmatic stage. The works of many of the Canadian artists profiled in this book can be usefully viewed in this premonitory sense, that of generating advance knowledge through the creation of artworks or systems that goes beyond their workings as technological objects as such or their aesthetic value.[53] For scholars and practitioners today concerned with reassessing the recklessness of disruptive innovation,[54] it is salutary to note that this kind of criticality combined with invention was often embedded within the actual creation and research practices of artists working during the earliest, formative period of digital media in Canada.

As "research-creation" becomes an ever more recognized watchword in higher education programs and funding schemes,[55] research policy can well stand to be better aware of closely documented historical cases that can serve as clear models and cautionary warnings. The book's central argument, that Canada's alternative technological ethos encompassed multiple ways of knowing, making, and valuing across the arts and technology, leading to multidimensional innovative outcomes, is put forward to help inform this kind of historically grounded policy formulation.

Northern Sparks

The cases in this book range broadly, both in terms of the arts and media practices covered, and in terms of the analytic resources needed to explain them. Whereas media art history has lately sought to become a platform for multiple media perspectives,[56] in practice most studies tend to bifurcate along conventional disciplinary lines between the visual and sonic realms.[57] This study, by contrast, spans film animation, music, the emergence of sound art and acoustic ecology, cybernetic cinema, interactive installation art, virtual reality, telecommunications art, software applications, and the branches reaching out from all of these creative practices to the emergent metadiscipline of human-computer interaction (HCI). To address such a breadth of practices, methodological approaches from several fields need to be enlisted, drawn in particular from the liminal space between the already interdisciplinary fields of media art history, and of innovation within science and technology policy studies. While some readers might find the level of close reading in some of these chapters too granular, others will no doubt lean in the opposite direction and want greater precision in my technical or artistic analysis. I have tried to strike the tricky balance found in the best sociologically oriented cultural criticism, of understanding "the works themselves" without losing sight of their historical and social contexts.[58]

This book is also informed by my own direct involvement in the media art and technology field in Canada. Between 1980 and 1993 I was employed at the Banff Centre for the Arts, where I was the founding director of its media arts division. From 1993 to 1997 I worked as a program manager for a Montréal-based federal research center for information technology innovation, founding a new program for "networked cultural information systems." In those same years, I also served as consulting policy analyst for new media at the Department of Canadian Heritage, within the directorate for arts policy.

My government service occurred immediately after the dissolution of the Department of Communications, which had since 1979 combined responsibilities for communication policy and research with arts and cultural development. Several overlapping policy exercises were going on at once, with chaos and bureaucratic infighting being par for the course. I often shuttled physically, as well as conceptually, between Heritage and Industry, trying to bring my practice-informed knowledge from the field to bear on the formulation of policy by the Information Highway Advisory Council, jobs-focused measures for new media training, and designing programs to support collaborative research between disciplines, arts centers, and research labs. This vantage point gave me a privileged view into the curious and often frustrating instability and incoherence Canada was experiencing during those years. As Marc Raboy and other scholars have documented,[59] when culture and communications policy in Canada was subjected to new globalizing pressures, important national histories were often overlooked or, at best, distorted. My own advocacy and analysis work for the Canadian government instigated exploratory research into some of the cases that are finally being published in this book.

The impetus for this book initially grew out of this experience as an active proponent of the arts in Canadian new media research, and therefore the methods employed in telling this history are in part ethnographic. Through my personal contacts, I was able to identify and interview many of the key figures in the cases studies, but in some instances, I also had direct knowledge as participant in or instigator of projects and initiatives that occurred within the time frame of the study. The interviews I carried out took the form of open-ended conversations, with individuals I selected as a researcher with an "inside observer" point of view. The names and dates of individual interviews are provided in Appendix A.

Chapter Summaries

This book begins with antecedents to and the deployment of some of the earliest applications of computing to filmmaking. Chapter 2 foregrounds Norman McLaren's work at the National Film Board, introducing an account of "proto-computationality" derived mainly from the procedural, notational, and quasi-algorithmic methods used in the films he produced from the late 1950s through the 1960s, mainly made with his closest collaborator at the NFB, Evelyn Lambart. McLaren's artisanal methods at the NFB are presented in detail, and they are shown to have exerted a powerful influence on the emerging computer culture in North America because of the way they illuminated the creative potential of computer graphics. The National Film Board as an innovative techno-artistic center is contrasted to the forces shaping the institutional development of computer graphics in the United States.

Chapter 3 presents a case of the co-development of art and technology via a sustained collaboration between the NFB and the National Research Council (NRC). In the mid-1960s, the NRC began a research program using interactive computing applied to "creative applications" in film animation and music composition. On commission from the NFB, the figure-based animation script of Peter Foldes's film *Hunger* was realized computationally by a team of NRC scientists and engineers, using a custom-built animation system supporting real-time direct manipulation modes of interaction between artist and computer. The research programs of the two sponsoring institutions were complementary, and the roles of the engineers, artists, producers, and programmers were fluidly defined and symbiotic. *Hunger* was the first computer animation to be nominated for an Academy Award (1974), and it won the grand prize for animation at the Cannes Film Festival. The chapter ends with an account of the later outcomes of this collaboration, beginning with a contentious incubation period when the NRC's system was used by young animators at the NFB. This usage contributed ultimately to the formation of one of Canada's early successful companies producing 3D animation software, Softimage, and the wider diffusion of the NRC-NFB nexus helped shape other successful business firms as well.

Chapter 4 presents a Canadian-oriented genealogy of artistically driven human-computer interaction, filling in some lacunae in the canonical histories of HCI, and pointing out the foreclosure of certain early potentials for interactivity in light of the standardization of graphical interfaces in personal computers in the later 1980s. The chapter begins by carrying forward the focus on animation from the previous two chapters and adding the results of research projects driven by applications in music, dance, and interactive installation art. The first example is a computer-mediated animation system using hand-drawn gestures first developed by Ron Baecker at the Massachusetts Institute of Technology's Lincoln Laboratory. Baecker carried this work forward in the 1970s at the University of Toronto's Department of Computer Science, which was early to support artistic exploration and development in both graphics and music. Joining Baecker at Toronto, musician Bill Buxton was influenced by the musical component of the same digital lab that had been used for computer animation at the National Research Council in Ottawa, as well as the previous analog electronic music instrument designs invented by Hugh Le Caine, also an NRC researcher, whose work since the 1940s anticipated the more widely known electronic synthesizers that appeared in the 1960s. Both of Buxton's prior influences emphasized the requirements of real-time performance and interaction at a time when these properties were in their infancy in computer music. Baecker and Buxton went on to become some of the leading figures in HCI as it emerged as a multidisciplinary field in the 1980s. Next the chapter considers Michael Snow's landscape film *La Région Centrale* (1971), which depicts a northern Québec landscape as seen through a custom-built mechanized camera eye. Importantly, the prescored program was replaced during the film's production by the artist's improvised gestures, but the programming concept was later implemented as a closed-circuit video installation (*De La*, 1971) at the

National Gallery of Canada. The codification of human gesture in dance is presented in an account of Lifeforms, the first widely used software application for computer-aided choreography, developed at Simon Fraser University in Vancouver by computer scientists and dancers, with input from and prominent application in the choreography of modern dance icon Merce Cunningham. The chapter closes with a consideration of Toronto media artist and programmer David Rokeby, whose *Very Nervous System* (1982–1991) uses a video tracking system that converts bodily movements into generative soundscapes. Alongside his landmark artistic advances, Rokeby was also an incisive critical theorist of configured interactivity, in writings that appeared just at the beginning of the euphoria about interactivity that spread rapidly with the mass diffusion of personal computers and the internet.

Chapter 5 pivots to performance, improvisation, and composition in music, which during these years in Canada enjoyed a remarkable surge in interdisciplinary energy and innovative outcomes. The chapter emphasizes the erosion of music's classical autonomy (initially formed in the nineteenth-century primacy of the romantic composer), the rising porosity of boundaries between music and other art forms, and the many interconnections drawn in Canada between music and other disciplines in the humanities, engineering, and social and natural sciences. This broader context is focused, first, through Glenn Gould's work as an author, which forecast a radical reconfiguration of the relationship between listener, composer, and performer, and also highlights his work as a creator of original radiophonic compositions that increasingly became the focus of his creative efforts after his retirement as a concert pianist. These original works blur the boundaries between music and language, documentary and fiction, and challenge listeners to expand their cognitive and perceptual capacities in ways that became formative for the future of sonic arts. The unique blend of free improvisation, sound experimentation, and popular folklore which comprises Québecois *musique actuelle* is highlighted in a discussion of various Canadian scenes for improvisational music, many drawing importantly on new audio and computer technologies. The alignment of the experimental energy of *musique actuelle* with the burgeoning nationalist forces within Québec society also occurred within an expanded aesthetic field blurring the formerly distinct boundaries separating music from other art forms. The chapter intensifies its focus on the Canadian contribution to the emancipation of music from its more calcified European classical moorings by recounting the origins and development of soundscape studies and the interdisciplinary field of acoustic ecology. The chapter also considers the creative and legal challenges to the concept of music as intellectual property put forward by the Toronto composer John Oswald, through his microsample-based composition and his genre-defining theorization of "plunderphonics." It concludes with a close reading of several of Gould's lesser-known "contrapuntal radio" documentaries, demonstrating how their sonic layering techniques conveyed his ambivalence about technological modernity.

Chapter 6 presents two contrasting faces of cultural networking. It examines the main case of instructive failure in the book, an account of Telidon, Canada's pre-internet digital network, a financial fiasco of the first order despite the enormous funding and prestige accorded to the project by government and industry. Cultural input to the system was belated, if still formative to the growth of new media culture in Canada; Telidon's government lab developers designed the system around an authoring protocol for both text and graphics, which was driven more by data compression imperatives than by a concrete understanding of how users might engage imaginatively with the system. The Telidon example is set against the contrasting development of a Canadian network of artist-run centers, emphasizing experiments in computer networking and real-time video experimentation. Whereas the top-down, bureaucratically driven imperatives of Telidon were unsuccessful, the bottom-up artist-driven experiments using telecommunication and art helped to foster an informal experience of networking and collaboration across vast distances, well before the advent of the World Wide Web. Among the key works analyzed in this section are a 1974 multisite performance by Vera Frenkel, *String Games*, that made early use of video conferencing technology, and a set of communication-art projects and exhibits in which Canadian artists were the international instigators of networking as a new expressive medium.

Chapter 7 presents a case study of the infancy of virtual reality (VR), based on activities undertaken between 1989 and 1993 at the Banff Centre in Alberta. It focuses on VR as an emblematic technology, considered both as an opportunity and a threat, initially as investigated critically by a diverse group of philosophers, engineers, and artists who met at the Banff Centre for a 10-week residency in 1991 on the "Bioapparatus."[60] During this residency, artists Catherine Richards and Nell Tenhaaf teamed up with the Banff Centre programming staff to investigate VR as a technology freshly open to artistic intervention, in order to take it beyond its initial scientific and military foundations, and to go beyond the quickly forming hype around the prospect of its widespread consumer adoption. At the core of the Bioapparatus construct is a notion of gendered embodiment in technology, and the chapter explores how the critical tenor of the residency elicited a rich techno-aesthetic controversy. The tensions inherent in the Bioapparatus residency were for the most part toned down in the next phase of the research program, the Art and Virtual Environments Project. This project attempted to envision various alternative futures for interactive immersive media environments by commissioning full productions of eight new artworks.[61] This was the earliest large-scale investigation of virtual reality as an artistic medium, and its lessons are little remembered—but useful to recall—during more recent revivals of VR.

Chapter 8: the book ends with a short description of how the coming decades would unfold, as Canada entered the 1990s facing a complex conjuncture of political, cultural, and economic challenges. In contrast to the initial generative confluence of the arts, media technology, and national consciousness presented at the outset of the book, the last decade of the century found Canada's response to the explosive growth of information and communication technologies to be partially anachronistic, containing a sometimes paradoxical mixture of heady enthusiasm and inward-looking retrenchment. The Canadian alternative technological ethos, as both creative method and interdisciplinary model, nonetheless continues into the present day, with the emphasis in Canada on media interface and interaction, the interpenetration of art and research, and the artist as creator of "counterenvironments." The Information Paradigm, that is, the wave of digital innovation that built during the 1970s and 1980s, and burst into a flood after 1992, may now be taken for granted everywhere, but the coming waves of innovation, still inchoate and emergent, are in just as urgent need of creative and critical intervention. The Canadian story serves as an instructive object lesson today and for the future, across the range of its broad scope and manifold methods for mobilizing the arts, policy, and innovation.

2 Norman McLaren and Proto-Computationality

The story of Canadian artistic and technological innovation recounted in this book begins in the heart of one its flagship state cultural institutions, the National Film Board. Founded in 1939 as the Government Film Commission, and operating as an independent Crown agency, its formal mandate from Parliament is "to produce and distribute and to promote the production and distribution of films designed to interpret Canada to Canadians and to other nations."[1] Following its early years, defined by an emphasis on public education and war propaganda films, after 1945 the NFB entered a period of sustained prominence in the production of technically innovative artistic animation. During these years, the NFB contributed in no small measure to the flourishing of experimental, auteur-driven animated filmmaking internationally, particularly in countries protected, as Canada was, from market forces by state patronage, such as the former Eastern bloc. Leveraging its close proximity to the United States, Canada also distributed its short animations through the mainstream industrial cinema from Hollywood, regularly providing the opening shorts before feature attractions released during this period. International festivals and prizes for independent animations proliferated, further nurturing the spirit of innovation and experimentation that spread from the NFB.[2]

These golden years of auteur animation emerged in part from the distinctive organization of production pioneered by the NFB's animation unit: work cultures emphasizing small teams, in contrast to the stratified hierarchic structures of industrial "orthodox" animation style typified by the Disney studio.[3] For animation theorist David Wells, orthodox animation stresses narrative form and is dialogue-driven, whereas experimental styles are more typically grounded in the inventive development of the materials of animation itself, and driven by a "dynamics of musicality."[4] The work of Norman McLaren typifies this emphasis on instrumental invention, musicality, and artisanal scale, and the NFB animation unit under his leadership developed a resourceful homegrown spirit that prided itself on being able to "put ourselves up against Hollywood with our own unique qualities," in the words of Colin Low, another of the Board's early leading lights.[5] This spirit had a political aspect as well, the contours of which are apparent in the NFB artists and technicians' active support for their American counterparts during the 1941 strike at Walt Disney Productions.[6] McLaren's close collaborator Evelyn Lambart recollected:

> We admired the work of [Disney striker] John Hubley and the European connection was close to our hearts. We loved anyone who was inventive, and we hated the fat cats. . . . Our work was exploratory and originality was valued tremendously. Derivative work was absolutely hated. We didn't do any cel work at all, in fact we were openly contemptuous of Disney.[7]

In short, the artisanal style of animation fostered at the NFB flourished after World War II within a relatively protected international scene that challenged, but also maintained close links to, US industrial production.

McLaren's work in particular was honored with high accolades, including a 1952 Academy Award for the anti-Cold War film *Neighbours*. In turn, his experimental ethos inspired some of the earliest American engineers and scientists in the emerging computer culture of the 1960s to turn their attention to filmmaking as a challenging arena for digital methods. The key aspects of a "computational sensibility" in the arts, according to digital artist and theorist Marc Downie, can be defined in two parts: a tendency toward generalization ("replacing constants with variables") and abstraction (the "re-expression of prototypes as templates").[8] Expressed in a less technical jargon, a computational visual sensibility is one which is able to apply formal rules in ways that can generate not just single images but a range of possible outcomes, transformations, or even new systems or rules. Another way to consider this sensibility for producing algorithmic images is in terms of procedures rather than objects. As articulated by Judson Rosebush, another digital artist and theoretician, a procedural approach to computation can be usefully distinguished by the use of computers just to simulate the look of classic painting styles, on one hand, or to automate the humdrum aspects of industrial production, like cel animation, on the other.[9] McLaren's work often exhibited precisely this kind of computational sensibility, particularly so during the 1950s and 1960s; not incidentally, these were the very decades when early ideas about computer-aided creativity were starting to circulate, often in the first instance within technical professional circles of computer engineers. In the words of one such engineer involved in early computer art, Ken Knowlton, McLaren's ideas about and the practice of animation at the NFB exerted a "magnetic attraction" on him and his colleagues. This chapter aims to understand this attractive force and place it into the context of the NFB's artisanal, anti-industrial-scale work culture. It will also analyze the "proto-computational" aspects of many of his films and their techniques, an analysis which insists that a computational sensibility can not only subsist within but thrive and be complementary to the artisanal, exploratory technological ethos typified by McLaren at the NFB.

Norman McLaren "Turns On" the Electrical Engineers[10]

Kar Liang

The year was 1966, and Kar Liang, an enthusiastic and inquisitive young electrical engineer at Northern Electric in Ottawa, struck up a casual conversation in the cafeteria of the National Research Council. The conversation changed the course of his life. Liang was a member of the small team at Northern Electric doing cutting-edge research on the computerization of telephone systems. In 1964, Northern had been acquired by Bell Canada, a long-brewing outcome arising out of a series of US court decrees beginning in 1949 ordering the divestiture of the AT&T monopoly. Liang's team still worked closely with AT&T's fabled Bell Labs, "like the little brother," he remembers; but Northern Electric's new business strategy was to break away from its scientifically subordinate role to the American telecom behemoth. Liang was part of an advance team at Northern—"they called us the 'suicide squad'"—surveying the world for the newest ideas in computer applications in telephony. In a pattern that will recur many times throughout this book, Liang's group was keen to absorb American scientific expertise and know-how, and it was also keen to apply this new expertise *for* Canada. Liang remembers the time with infectious enthusiasm, since the team enjoyed a vital *esprit de corps*: be daring, be creative, and don't be afraid to talk things over.

The NFB shared a cafeteria with the National Research Council, the flagship national research laboratory. So when Liang heard from Gerry Graham, head of research and technical development at the National Film

Board of Canada, that Graham would be presenting a program of films by Norman McLaren later that afternoon at the NRC, he thought, why not? Liang had never heard of McLaren, but as an avid amateur photographer, familiar with darkrooms and optics, he was intrigued by Graham's description of the Scottish-born Canadian animator's style.

Liang responded immediately to McLaren's innovative technical approaches, as well as the creative quality and scope of his imagination. "I was captured," he recalls. "What intrigued me the most was the one he did without camera, painted on film, *Begone Dull Care*." Liang realized that in normal use, the camera is an intermediate medium, transforming imagery through chemical reactions. "But if an artist can convey his feeling, his insight, his emotion, idea, directly onto the medium itself, the effect ought to be more direct. I was touched by the immediacy. Also, I felt there was an interplay, between the imagery and sound, that also had a unique character I'd never seen in other movies." Liang was so enthralled that, after expressing his enthusiasm to Graham, he was invited to visit the Film Board and meet McLaren in person.

> I toured the whole plant, was introduced to McLaren and the other animators, and also the documentary producers. The whole place was like a big system, with various processes, steps, and operations. I was very impressed with the kind of ambience I found in the place. It just seemed to emanate a feeling of creativity, from all these people, creating and producing images. I noticed an interesting kind of rapport among them, how they worked as a team.

It was evident to Liang that "Norman was showing the way, with his creativity, working with camera, without it, and always trying out new things. I could see it as a working laboratory, complex yet organized." He viewed the research in progress for the NFB's multiscreen project, *Labyrinth*, that would occupy a pavilion at Expo 67 in Montréal the following year. He was also impressed by the documentary division, pioneering at this time its Challenge for Change program, which helped him gain a new realization of the social role film can play in explaining complex subjects to the general public.

Liang joined the NFB the next year to start work on a computer-controlled animation stand. "I said to Gerry, to be successful, let me learn about all aspects of filmmaking, including doing it; and give me a budget to travel, so I can learn from the people who have already done lots of work in this area or related areas." There was little money to support new initiatives; Liang inherited a lesson about resourcefulness instilled by McLaren: "the less money you have, the more imagination you need." His first assignment was to test new techniques for sound synchronization, for which he produced his own short film. Liang also gathered intelligence broadly, just as he had done at Northern Electric. With his connections to Bell Labs, he soon became a fixture among the US researchers working in computer graphics, through an early user group sharing a common interest in automated information display equipment. "I wasn't a filmmaker myself, but I was the only one in those meetings who really understood film production, from the inside. The rest came from the military, from industrial research, but they all respected the technical innovations of the NFB."

He was also quick to realize how the Film Board animators, guided by McLaren's own creative process, tended to refuse ready-made solutions to problems. "When I got there, they showed me a memo from the early 1960s, from Bell Labs in fact, that was supposed to lay out all the solutions for animation, using analog electronics." Liang recalls how the animators categorically rejected the well-intentioned memo: "All the animators, those on the creative side, don't like to accept technology as a given—'this is something you can use, to improve your operation.'" McLaren's filmmaking credo, published decades earlier, had precisely articulated their creative approach:

1) attempting to keep at a minimum the technical mechanism standing between my conception and the finished work.
2) handling personally the mechanisms that do remain, in as intimate way as a painter her painting, or a violinist his violin.
3) making the very limitations of these mechanisms, when brought in touch with the theme, the growing point for visual ideas.
4) making sure of a chance for improvisation at the moment of shooting or drawing.[11]

Liang would only stay at the NFB for a few years but accomplished a great deal in that short time. He led the team that produced one of the first computer-controlled animation stands, instigated contacts at the National Research Council that led to a germinal production collaboration with the NFB (see chapter 3), and even gave McLaren the chance to make an animated computer film himself, *Birdlings*, which Liang showed to technical groups across the continent. "That film caused a sensation, even though it was so simple, just sticks moving, compared to all the very sophisticated work the computer graphics people were trying to do." For Liang, McLaren's test computer film offered a clear demonstration of McLaren's oft-quoted definition of animation, as "not the art of drawings that move, but of movements that are drawn. What happens between each frame is more important than what exists on each frame. Animation is therefore the art of manipulating the invisible interstices that lie between frames."[12]

Kar Liang's role as an intermediary, between the NFB and the wider North American research community, or more broadly as a boundary spanner between art and engineering, would continue in a series of postings in the federal government.[13]

Greg Hermanovic

Stationed on a Canadian navy weather research vessel off the coast of Senegal in 1973, Greg Hermanovic was programming a mainframe computer to play Pong. The Montréal native, recently graduated in mathematics and physics from McGill University, had time on his hands, and the computer, a DEC PDP-11, was only being used twice a day to take brief measurements. His Pong game became something of a hit when Russian generals boarded the ship and were invited to play in "the first international computer game tournament." Hermanovic recalls that his interest had always been to build tools, but that he had gotten bored with "the engineering thing . . . I wanted something a bit more loose and open, as opposed to building systems others specified. And I always liked things that involved motion—sports, music, simulation. Computer graphics, animation is motion. Building tools for artists is a natural for me."

During high school at Dawson College in Montréal, a school known for its free-form classes and field trips to artists' lofts in New York City, Hermanovic attended a series of experimental film festivals as a teenager. "McLaren's films were mind-blowing. Some are mechanical, almost programmable. Like the film with color bars, *Synchromy*, synched with hand-drawn sound. Or *Mosaic*, which I saw over and over, with the mechanical nature of its moving dots. There was really a wide body of work shown at Dawson in those years, Jordan Bellson, John Whitney, Stan Brakhage, Michael Snow. This was the big push for me, as far as understanding visual things."

After a stint programming flight simulators at Canadian Aerospace and Electronics, Hermanovic took time off work following an injury. "I made a lot of pictures and lists and diagrams of how I would use a digital computer to redo McLaren's *Synchromy*, which I was showing (among other strange stuff) in micro-film festivals at my country place. At that time I was thinking how to program some of McLaren's films." An avid tinkerer with modular music synthesizers, he went on to combine an interest in real-time controllers and patchable electronic music systems with computer graphics. In 1987, he cofounded Side Effects, a Toronto-based software company offering a unique procedural tool set to animators, for which he would eventually win several Academy Awards for Scientific and Technical Achievement in the 1990s.

Ron Baecker

By the mid-1960s, the Massachusetts Institute of Technology's Lincoln Laboratory had become the epicenter of real-time computer graphics research. Out of a technological foundation developed for Cold War air missile defense, a set of unprecedented graphic interaction techniques were developed.[14] Among the most influential was Ivan Sutherland's Sketchpad, "an interactive design tool for the creation, manipulation, and display of geometric objects in two or three-dimensional space. The system permitted operators to sketch with a light pen on the face of the monitor, change the position and size of objects, duplicate objects, and paste them into an evolving design."[15]

One of the electrical engineering PhD students working on computer graphics at Lincoln Lab was Ron Baecker, a film enthusiast whose 1969 dissertation demonstrated a new way to use computers for "interactive computer mediated animation." Baecker was spending his spare time hanging out at Harvard's Carpenter Center for the Visual Arts, where he was impressed by the range of experimental techniques and creativity found in the films of the NFB. In McLaren's work he saw "incredible richness of motion control; simple motion over complexity of imagery." Baecker quoted McLaren's definition of animation in his dissertation, as "not the art of drawings that move but the art of movements that are drawn."[16] Baecker recalls how McLaren started him thinking: "What could you do to couple an artist directly to the computer?" He invited a Harvard animation instructor, Eric Martin, to collaborate with him on a method for drawing motion directly into the TX-2. This would become his Genesys system, shown widely across the United States at computer conferences, and whose originality would land Baecker an assistant professorship at the University of Toronto. In Toronto, Baecker found a fertile creative scene, and a politically conducive home for a series of further advances on his initial animation research. He would become a founding figure in the art and science of human-computer interaction, a story that appears in more detail in chapter 4 below.

Ken Knowlton

Parallel to Baecker's artist-oriented research on animation at Lincoln Lab, the Bell Laboratories in Murray Hill, New Jersey also sponsored a sustained artistic research program exploring new applications for computing. The Communication Science Division at Bell Labs was the site of key innovations of the information age, notably the transistor, satellites, and information theory. The earliest languages for computer music and animation were developed there by 1965. Research engineer Ken Knowlton was author of the BEFLIX animation language, which treated the entire viewing area as a rectangular array of dots. The language included powerful expressions for the use of algorithms to generate image patterns. Within a research culture conducive to collaboration with artists (this was during the same years laser scientist Billy Klüver cofounded Experiments in Art and Technology out of Bell Labs), Knowlton began working with filmmaker Stan VanDerBeek in 1966. Together, they went on to co-create a series of nine experimental films, the *Poemfields* series, that were widely exhibited and are now recognized as significant milestones in computer animation.[17]

Knowlton was well acquainted with the work of the National Film Board, having worked closely with Kar Liang during the latter's tenure as research liaison officer. He visited Montréal for Expo 67, and while there got to know McLaren well. Like Liang, he was initially fascinated by the animator's work drawing directly on the film. "I was very impressed by the NFB. I considered coming up to Canada, learning French. It was the magnetic attraction of the NFB. What impressed me? McLaren." The atmosphere around the NFB was so compelling that he applied and received approval for a Bell Labs-funded sabbatical year in Montréal in 1968–1969.

> So why didn't I move to Montréal? Martin Luther King was assassinated, there were race riots in our home town and my wife and I got involved in politics as a race-relations peacemaker. We were suddenly inextricably involved in local affairs and the Montréal idea was shifted to the back burner and it finally but slowly evaporated.

Knowlton stuck around Bell Labs for another decade, but his heart had changed and his career path led him to become an artist himself. By then AT&T was about to be split up by court order and the research climate tightened. During the 1960s, Knowlton recalls Bell Labs as a free-flowing environment. "There wasn't really any strong policy at Bell Labs at the beginning. If you were doing something remotely interesting, you could keep doing your thing. There was a general interest in learning about vision, hearing, perceptual processing.

By the late 1970s, it was becoming clear that deregulation and competition would change Bell Labs."[18] Widely honored as one of the most important pioneers of computer graphics, he has spent the remainder of his career as a visual artist, specializing in mosaic-based portraiture using seashells.

Taken together, these biographical sketches reveal the way the emerging computer culture of the 1960s would find ways to pursue expressive possibilities growing out of two distinct, only apparently disparate sources. First was the obvious inherent sensibility of computation itself—its grounding in procedural logic, algorithmic thinking, and systematic transformation; second was the loose, improvisatory, restlessly innovative direct hands-on work style of the experimental animator, typified in the personality and oeuvre of Norman McLaren.

Understanding the Magnetic Attraction of Norman McLaren

> If I were a young artist, I would be using computers to make a new kind animation, a new kind of film. That's right and proper, as it should be. If I were growing up now, I would have gone right in and tried to get hold of a computer. But one is the child of a certain age, born at a certain time; one uses certain things.[19]

The plain facts of Norman McLaren's life can be narrated simply, as befits a man who spent most of it employed, not always easily, as a civil servant. He was born in Stirling, Scotland, in 1914. His earliest artistic experiences were musical, and he received six years of formal training in piano and violin. A synesthete, he discovered as a teenager how cinema might become a means to express this synesthesia. His first animation work occurred at the British Government Post Office Film Unit, where he was spotted as a coming talent by another Scot, John Grierson, the GPO Film Unit chief. When Grierson was appointed Canada's Government Film Commissioner, he invited McLaren to establish the Film Board's animation department. As a government employee, McLaren received the kind of steady, calm and consistent support that matched his temperament and always delicate health. He stayed there until his retirement in 1980, producing, like clockwork, a steady stream of artistic and educational films, technical reports, and experimental test films—explorations of new techniques that he judged to have failed and were never released. The gruff and demanding Grierson later remarked that McLaren was "the most protected artist in the history of the cinema. No suffering; that makes him suffer."[20]

In this way, McLaren is an unusual character in the history of the experimental arts: a state artist, or *artiste d'état*. His temperament distanced him from the dissident, transgressive ethos of the historical avant-garde, and he once jokingly remarked that he would rather call himself an artist of the "*arrière-garde*," not in reactionary terms, but in the way he chose to avoid any adhesion to movements, factions, or cliques.[21] He saw his work as a continuous systematization of the discoveries and intuitions of others, all through the prism of a disciplined working method grounded in the materiality of cinema—making cinema "image by image," meticulously and with a rigorous precision that found the inspiration for each film in the invention of new techniques.

However, the years of McLaren's youth and early adulthood were also marked by the political upheavals of the Depression and 1930s. His path was parallel to that taken by many socially committed artists who came of age in those strife-filled decades. A leftist from his boyhood years, like many fellow-travelers in Communist cultural circles he visited the Soviet Union, before the Stalin show trials, and volunteered as a cameraman during the Spanish Civil War in 1936. He was a lifelong pacifist, even during World War II, but nonetheless contributed to the British and Canadian war propaganda efforts, owing to his revulsion toward Hitler's racial program. A genial worker and inveterate collaborator, shy yet deeply considerate and polite, McLaren found his wartime service bearable, despite his reluctance: "I made light films with war messages."[22]

The early 1940s saw McLaren in New York, immersed in the city's bustling experimental and abstract film culture. It was in New York that he perfected his own version of hand-drawn cinema that would reach its first apogee in Ottawa in 1949, with *Begone Dull Care*. During the late 1940s, he experienced what he considered the major interregnum of his four-decade career in Canada, when he took leave from the NFB to travel in China and India. "Even though I was only there for the beginning of the revolution of Mao, my faith in humanity was restored."[23]

Throughout the period under close examination here, the 1950s–1960s, McLaren's closest collaborator was Evelyn Lambart, one of the few female animators at the NFB. An archival photograph shows the two artists in close collaboration, a common mind expressed between the two pairs of hands working in synchrony (figure 2.1). Lambart was one of the NFB's earliest employees, a key team member working on a wide range

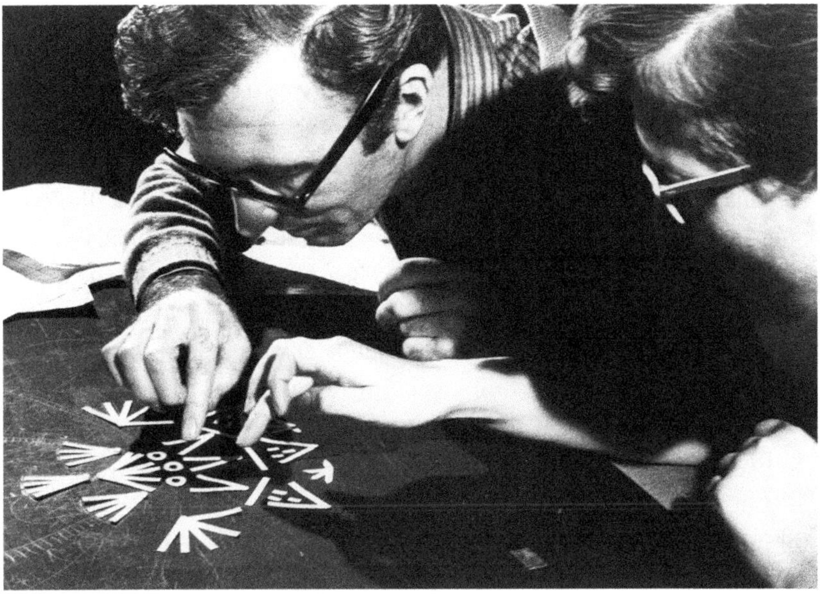

Figure 2.1
Norman McLaren with Evelyn Lambart around 1958.

of projects, and noted for her precision and draftsmanship; the formal rigor and clarity of her organizational style enabled a close synergy with McLaren's more spontaneous and restless ingenuity. An independent spirit, Lambart thrived in the Film Board's unique work culture. When McLaren shifted his attention in the late 1960s to dance films, renewing a lifelong passion for the ballet, Lambart left his side, having been his most important collaborator for almost 30 years. The dance work held no interest to her, so she went on to make her own films for the remainder of her career; she retired in 1974.

McLaren scholarship is extensive. Yet recent authors have made surprisingly little use of two of the most useful studies of McLaren's work for understanding the artist's relation to emerging technologies during the 1960s. "Le mystère d'un cinéma instrumental," by André Martin, appeared over several issues of the journal *Cahiers du Cinéma* in 1958; Alfio Bastiancich's book *Norman McLaren: Précurseur des nouvelles images* was published in Italian in 1981 and translated into French in 1997.[24] Together, these two accounts of McLaren's work set up the following consideration of McLaren's and Lambart's combined approach to experimental filmmaking informed by a proto-computational sensibility.

André Martin is the first of several key French connections that will appear in the course of this book, owing to the well-traveled historical circuits between France and Québec. A member of the influential Cahiers du Cinéma group in Paris, Martin was, like Truffaut and Godard, a film critic and cinephile, who hailed from Bordeaux. His attraction to Canada was initially sparked by the creativity of the NFB, as well as the media-theoretical writings of McLuhan and Harold Innis. He arrived in Canada in the mid-1960s and stayed for a decade.[25] He would later write one of the most well-informed articles on early computer animation practices in 1976, appearing in the Cinémathèque Québecois journal *Cinéma/Québec*.[26]

To a seasoned French film critic in 1958, what was the mystery of an "instrumental" cinema, as Martin's article called it? The answer to this question will shed light on the specific interest in McLaren evinced by the early computer engineers. The first principle is what McLaren himself called "hand-made" cinema, whether this meant making inscriptions directly on film, manipulation of materials and paint, manipulation of cut-outs, or so-called "pixilation," the use of stop-motion photography (such as in the Oscar-winning short *Neighbours*).

Second is what Martin called "notation," a precise particularization of visual elements as a composable vocabulary, so as to enable their organization like a musical score. As a rule, Lambart would handle the notation schemes invented for each film, using dope sheets, allowing McLaren, the more intuitive artist, to play with options much as a composer manipulates phrases and motifs. While not a serialist in formal terms (though the musical twelve-tone school was part of McLaren's imaginative landscape), he made extensive use of serialist methods, like inversion, transposition, and retrograde motion, as well as classical counterpoint, like canon and the fugue.

The third element of the "mystery" for Martin was systematic research, the continuous production of prototypes and études, with limited release to professionals, and the circulation of technical reports. The technique of hand-drawn sound, for example, was developed with Lambart initially in a test film entitled *Workshop Experiment in Animated Sound* in 1949. In Martin's words:

> *Workshop Experiments* is not a film for general distribution, but rather a memento intended for use by artists interested in new methods of music composition and sound effects. The film also recounts the initial steps needed to establish the basis for a supple technique.[27]

Manual instrumentality, notationality, prototypicality: according to Martin, these concepts together serve to demystify McLaren's ostensible magic. Relying heavily on Martin's articles, Italian critic Bastiancich carried this analysis forward in 1981, covering McLaren's output well into the early years of electronic and digital image-making—what are called in the French translation "nouvelles images." For Bastiancich, the stakes in the transition were epochal. Moving from analog images (photographic and electronic) to computer-generated images, he suggested, is comparable to the transition in 1880–1895 from the magic lantern to cinema.[28] With his method of making films frame by frame, McLaren appeared to him "an artist capable of mastering digital images by giving them a poetic aspect putting the technology in second place, leaving the artistic aspect to flourish."[29]

Image-by-image cinema is "intimately connected with technical research," and the "natural adversary of the passive usage of the camera, ever since we started to use it as a means of reproducing reality."[30] For Bastiancich, it is Meliès, not the Lumière brothers, who stood at the headwaters of the stream leading to McLaren. "Like painters, sculptors, and poets," Bastiancich writes, McLaren "does not limit himself to working with materials using existing techniques, but rather carries forward his work in a constant stylistic and formal evolution, with a poetic autonomy that belongs uniquely to the artist, and not the artisan."[31] Paraphrasing Christian Metz, he remarks that this branch of the experimental animation tradition provides the real alternative to literary or narrative cinema, the "trait d'union entre l'art cinématographique et l'art pictural" (the link between cinema and pictorial art).[32]

All this depends on systematic research, a research looking for a matching of technique and expression. McLaren scholar Terence Dobson echoes the same view, claiming that in McLaren's work the "relationship between technique and art was symbiotic."[33] Citing Martin, Bastiancich underlines how McLaren's freedom of expression belongs to the avant-garde cinema, but to an avant-garde in which the "nonconformity is not belligerent, and which sets out to make a revolution without sacrilege or aggression."[34]

Many years later, Pierre Hébert, one of McLaren's closest protégés and colleagues at the NFB, would publish a treatise in his retirement devoted in part to understanding his life's work and tutelage under McLaren. He expressed his credo and his debt to McLaren's lessons in these terms:

> To keep doing animated films, I need to give the practice the form of an allegory that permits me to handle the apparatus as meaning, in such a way that the manner in which I use it to produce technical effects constitutes in itself a mode of signification.[35]

The Experimental Films and Études of the 1960s

This section considers the diverse production of McLaren and his close collaborators, including Lambart and others, with the overall aim of recontextualizing this work as proto-computational.[36] The diversity of the output from the NFB's animation unit during McLaren's direction is astounding: a group of camera-less films produced directly on film stock; chronophotographic productions, using overlaid phasing of sequences of filmed body poses in dance films; several visualizations of musical counterpoint; an unrealized flicker film experimenting with nonphotographic light modulation; tests of, and a follow-up documentary about, pinscreen animation (considered here as a form of mechanical pixel graphics, or "analog framebuffer"); and—not least—a film realized by McLaren using a real-time computer system, which served just as another experimental test and was never released in his own lifetime.[37]

The *Lines* Trilogy

Turning to a reading of McLaren's films with a mind to their proto-computationality, it is well to begin with a trio of works all co-produced with Evelyn Lambart, and which were derived from the same motion dynamics: *Vertical Lines* (1959), *Horizontal Lines* (1960), and *Mosaic* (1965). As McLaren commented on their working relationship, which was particularly close in these works, "without Eve, the majority of my films before 1965 could never have been produced. Incredibly meticulous, she excelled in the organizational aspects, while being a fine artist in her own right. Once my idea had been determined, we worked together to find practical solutions, in which domain she was much better than me."[38] In the case of the first of the *Lines* trilogy, *Vertical Lines*, the task was to limit all imagery to verticals made by etching on 35 mm black leader. To carry out the exacting task perfectly, special steel rulers were ordered from England, and extensive testing carried out with the sharpness of knives to allow variation in thickness. Imperfections in the emulsion produced ragged edges and were kept in the final film. The white lines appear against changing color backgrounds moving from unicity—a single stationary line—to multiplicity with an accelerating density to a sudden pause, followed by numerical reduction to a single line again. Width was also used to a powerful aesthetic end, where lines came together to produce variable thicknesses, which at times occasions a sense of three-dimensional perspective. In the words of painter Hans Hoffmann, in terms that well fit the use of line in these films, "The width of a line may present the idea of infinity. . . . Tension and movement, or movement and counter-movement, lawfully ordered within unity, paralleling the artist's life-experience and his artistic and human discipline, endow the work with the power to stir the observer rhythmically to a response to living, spiritual totality."[39] Aimée Mollaghan's essay on these films, from which the preceding quote is drawn, points to several ways in which McLaren anticipated minimalism in visual arts and music, including serial patterning in the addition and subtraction of unitary elements or cells. Simple arithmetic progressions like these were extended with powerful elegance and utmost simplicity the following year, when Lambart and McLaren contrived to flip the vertical lines 90 degrees. They began by viewing the first film with their "heads cocked horizontally," allowing them to see that there was "something more to the effect than a mere change in the angle of the lines. In fact, with horizontal lines, our minds read into their rising and falling motion the existence of gravity. Since the lines did not decelerate as they rose up, nor accelerate as they fell down, they seemed to float. This seemed sufficient reason to have our entire vertical lines turned on its side."[40] As David Curtis has noted, "the lines suggested procedures," and—seeking to align McLaren with modernist tendencies toward formalist rigor—he goes on to state that "[McLaren] is at last prepared to adopt them [procedures] without compromise."[41]

The calculated, formal rigor of which Curtis is so approving took a further step five years later in *Mosaic* (1965), which consists almost wholly of the *intersection* between the two previous *Lines* films, made by "running a clear-on-black copy of both . . . in contact with each other, in an optical printer."[42] The soundtrack was produced by engraving marks on the black emulsion-coated leader, as will be discussed in detail in the following section. Various accounts exist of the motivation behind combining the two *Lines* films by deriving their intersecting points. These remaining points produce moving matrices that quite literally instantiate the phrase used by McLaren to define the essential artistic element of animated filmmaking, by visualizing the "invisible interstices" between objects in flux.[43]

One story about *Mosaic*, possibly apocryphal, is of serendipitous discovery. When the two prior films were accidentally projected on the same screen simultaneously, visiting British mathematician Paul Briggs "swooned" over the elegant result, calling it the pinnacle of abstraction, "stripped to its bare essentials."[44] As previously noted, Greg Hermanovic took it as a challenge to calculate an emulated version of these three films computationally, and pianist Glenn Gould was similarly drawn to make a comparison with musical techniques—in this case not with minimalist composition, but with Schoenberg's serial operations of inversion and transposition.[45] Curtis, the formalist critic, was critical of the opening of *Mosaic*, in which McLaren himself appears on camera casting off a particle into space, as if to "endow it with anthropomorphic life," which for Curtis only served to "mystify his procedure."[46] Rather than seeing this introduction, and the symmetrical conclusion in which McLaren comically picks up the sole remaining point, as a sign of formal weakness, one may see it as illustrative of the filmmaker's desire to avoid what he considered "too esoteric" outcomes in even his most abstract expressions.

Experiments with *Flicker*

During the Second World War, McLaren produced a propaganda film enjoining Canadians to "Keep their Mouths Shut" that began with a terrifying sequence of a skull seen against a burst of stroboscopic flashes. The alternation of black-white frames, made with George Dunning, another early NFB animator (who went on to future fame as the director of *Yellow Submarine*), lasts but a few seconds, but it planted a seed in McLaren's mind that came back fifteen years later when he was asked to take part in a scientific film about the brain. Asked by researchers at McGill University's Neurological Institute to visualize the various rates of brain waves, McLaren came back to the wartime flashing klieg light technique; while the film on the brain was never made, he produced but never completed an astonishing three-minute test, again with hand-drawn sound, called *Flicker*.

McLaren's work with *Flicker* precedes by five years the influential film *The Flicker* by Tony Conrad, which has received much more extensive art historical as well as philosophical attention. Whereas Conrad's work, as recounted by Branden Joseph, had roots both in the literary and musical avant-gardes,[47] McLaren's came out of war propaganda and scientific illustration. Both spatialize time in profoundly intensive ways, by meticulously composing rhythmic cycles of what is essentially a binary code of alternating black and white images.

Gilles Deleuze, in his second volume on the philosophy of cinema, dealing with the time-image, took Conrad's *The Flicker* to inaugurate nothing less than a new era in abstract or "eidetic" cinema. Acknowledging McLaren's role in also inventing a cameraless cinema, directly producing not only images but also sounds, Deleuze singled out Conrad's film as imparting not the "filmic process" itself, but "a cerebral process. A flickering brain, which relinks or creates loops—this is cinema . . . a cinema of expansion without camera, and also without screen or film stock."[48]

Yet there is a key difference between McLaren's and Conrad's flicker films: the later American artist only used black and white, whereas McLaren included some gray frames as well. As Donald McWilliam points out, this allows McLaren to produce a series of "overlapping fades [from] black to white, white to black." Asked why he never finished and released the film, McLaren's response was that it was "too esoteric, even for me."[49] Yet he remained fascinated with and enthusiastic about flicker, and was delighted by his close protégé René Jodoin's *Rectangles et rectangles* (1980), which he called the "ultimate flicker film."[50] Not incidentally for a discussion of McLaren as a crucial proto-computational filmmaker, Jodoin was the first head of the NFB's French animation division, which became the institutional crucible for the advancement of the experimental artisanal ethos of McLaren after its founding in the mid-1960s. (The English animation unit moved in large part toward conventional, cel-based drawn animation styles.) Jodoin's *Rectangles* was produced using the NFB's first computer animation system, developed between 1968 and 1973, which will be main technical subject of the next chapter of this book.

Animating Sound: Hand-Drawn Music from the 1940s to *Synchromy* (1971)

McLaren was exposed as a youth in the UK to European techniques for drawing sound,[51] and began producing his own soundtracks this way in the 1940s. A document of his own extensive techniques was released for professional educational distribution in 1949, as previously mentioned, under the title "Workshop Experiment." McLaren painted a set of cards, covering the six octaves of an equally tempered scale, and photographed them for printing on the optical soundtrack. McLaren, the "*arrière-garde*" artist, was uninterested in pushing his technique into more venturesome musical directions. (John Cage, who was familiar with it in the 1950s, encouraged McLaren personally to use this directly hand-painted method in order produce microtonal frequencies, instead of conventional melodies, but McLaren did not follow the suggestion.)[52] As noted above, *Mosaic*'s soundtrack was produced using this method, but it reached its ultimate realization in 1971 with *Synchromy*—one of the rare McLaren films produced without Evelyn Lambart's collaboration. This is widely considered the apotheosis of synesthetic cinema, if only in the strictly technical sense that the animated soundtrack consists of exactly the same patterns of graphic shapes that comprise the visual track. In literal terms, "what you see is what you hear." Again, in animating sound directly on film, McLaren was consolidating and improving techniques he did not invent, but which he had mastered and improved upon. *Synchromy* was shot frame by frame using the 72 cards each with "stripes and striations, all representing a musical semitone higher or lower. The more stripes the higher the note, the fewer stripes, the deeper the note."[53] The music was composed first, then filmed and colorized in the optical printing process.

Two test films for *Synchromy* exist, outlining the systematic process with which McLaren plotted out rhythmic and metric distributions of sound. The final version begins with a didactic exposition of the image-sound relationship, note by note, then in phrases, and differentiating by width the loudness and by density the frequency of each sound. The musical content is lively, a sprightly boogie-woogie that eventually becomes a frenzied blur. The McLaren scholar Terence Dobson analyzes the audiovisual innovation of this film astutely:

> Seeing and hearing operate in different realms. Their perceptual processes differ from each other. The most common form of musical perception is based on the recognition of short groups of notes in a phrase. In *Synchromy*, despite the fact that each of McLaren's notes lasts just a fraction of a second and is interspersed with micro silences, the listener is able to discern structured phrases. However, the visual perception on screen of these same brief pulses and their accompanying micro-silences differs. Visually, the manifestation of the micro-silences is perceived as successive temporal gaps. The result is the perception of a flicker.[54]

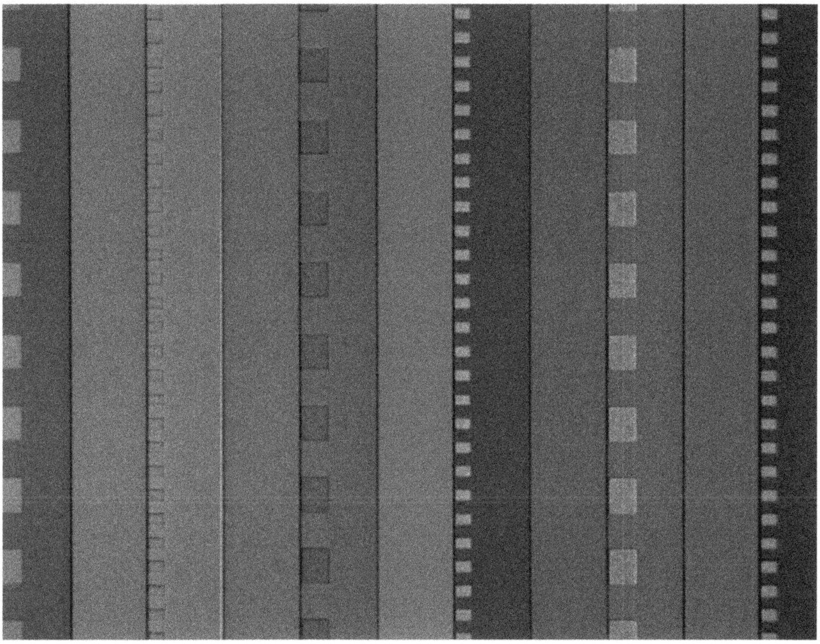

Figure 2.2
Norman McLaren, *Synchrony*, 1971.

In other words, the sonic message can easily be grouped as melody, whereas the visual effect is more visceral, growing as the flashing light increases in rate and intensity. For Dobson this produces an inherent tension, in which the initial recognition of the synesthetic relationship veers from "knowing to unknowing, from expectation to shock, excitement to joy."[55] This is a textbook "transcoding" of graphic to sonic shapes; it should be noted, indeed, that for digital art theorist Lev Manovich, this operation more generally constitutes the deepest and most far-reaching of his attempted codification of five "principles of new media."[56] McLaren's *Synchromy* is an essentially "proto-computational" work in its laboriously hand-crafted method of realizing an operation of multimodal translation, which later would become a routinized, automated core component of digital grammar.

Pinscreen Tests and Documentary Film

In its role as a state cultural institution, and with strong links to international independent animation production, the NFB often hosted residencies and commissioned artists from abroad for educational workshops.

The French animator Alexandre Alexeieff and his collaborator Claire Parker were such visitors in 1973, bringing to Montréal a unique animation technique they developed using 240,000 metal pins arrayed on a frame. Lighting the pin patterns produces differential shadows according to the extent of their protrusion from the screen. McLaren helped promote and spread information about pinscreen animation, supporting the couple's work by sponsoring residencies and by translating Alexeieff's theoretical writings. For present purposes, it is useful to consider the technique as a form of proto-computational pixel graphics, in which the discrete points indicated by each pin are modulated variably, therefore serving well to produce malleable shapes conducive to continuous transformations. In addition to his role supporting Alexeieff and Parker, which concluded with a 35-minute documentary, *Pinscreen*, McLaren himself produced a one-minute pinscreen test film in 1961. This was made using Alexeieff's apparatus in Paris. The test was abandoned, as McLaren declared that even his own notable patience was exceeded by the demands of the pinscreen system. The test film contains tantalizing glimpses only, such as variably accelerating swirls and swoops of circles, dots, and squares, in the manner of Oscar Fischinger's numbered *Studies* from the 1930s; morphing of points to circles to sunflowers; and a zooming circle that appears to advance through the screen. In proto-computational terms, it is notable that the first end-user programming language for computer animation, written by Ken Knowlton at Bell Labs, also was a bitmap-based method of coding, rather than one based in vectors or geometry. (Knowlton's attraction to McLaren's cameraless animation and direct instrumental techniques has been noted above.)

Musical Polyphony: *Canon* and *Spheres* (with René Jodoin)

Canon (1964) and *Spheres* (1969) are music-driven animations whose purpose is the visualization of polyphonic textures. *Canon* is in three parts, showing the contrasting ways in which phased or displaced temporal relationships between elements can be represented: the first uses physical blocks, the second 2D cutouts, and the third is a stop-motion animation of human characters, in the manner of McLaren's *Neighbours*. By presenting the idea of musical canon in contrasting ways, McLaren shows how the concept is independent of the material itself. *Spheres*, a collaboration with Jodoin, has a long lineage that illustrates once again the way McLaren worked as a research artist, drawing on a reference library of materials. The initial idea for *Spheres* emerged out of a commissioned film about nuclear energy from the 1940s, called *The Chalk River Ballet*. Chalk River, located near Ottawa, is the site of an early nuclear laboratory, first supporting the British atomic research program and later serving as the research basis for the manufacture of Canadian nuclear reactors. The initial film was to be a scientific illustration of nuclear energy and was meant to have a drawn soundtrack. The work was set aside, according to Jodoin, after a pair of nuclear explosions at Chalk River, which produced serious radioactive leakage. The same material was revived and completed in the late 1960s, using the music of Bach as performed by Glenn Gould. (There remains only a vestigial reference to the atomic energy origin of the project.)

Pas de Deux

In *Pas de Deux*, McLaren filmed a male and female dancer in varying poses, using overlapping multiple exposures. Again, he developed the project over several decades, taking his initial inspiration from a French advertising film that presented a woman running in slow motion displaying the multiple overlaid phases of her actions, in the manner of Étienne-Jules Marey's chronophotography.[57] Working from this initial source material, he performed numerous tests of the concept of stagger and delay, varying the number of delayed images, the tempi of their trailing paths, and contrasting diverse movement and choreographic styles. Once again, a key expressive feature is the multiplication by manual means of the body's contour along a compositional continuum of up to 35 trailing images. The production of these trailing afterimages required laborious optical compositing, in a manner which can easily be automated today with software such as Adobe's After Effects. As a proto-computational film, addressing with great purity the variation in what can be considered the "temporal aperture" of experience, *Pas de Deux* has inspired several recent critical commentaries. For movement scholar Erin Manning, "in *Pas de Deux*, McLaren uses a chronophotographic apparatus to focus not on movement's poses but on the activity of the interval through which movement's preacceleration can be felt. *Pas de Deux* is an experience of grace taking form."[58] Allana Thain considers McLaren's body of work at large as a "precursor to new media, already engaging with and challenging many of the distinctions that have been key areas of debate in cinema studies following the mainstreaming of digital technologies."[59] These two views on *Pas de Deux* reinforce the general claim being advanced

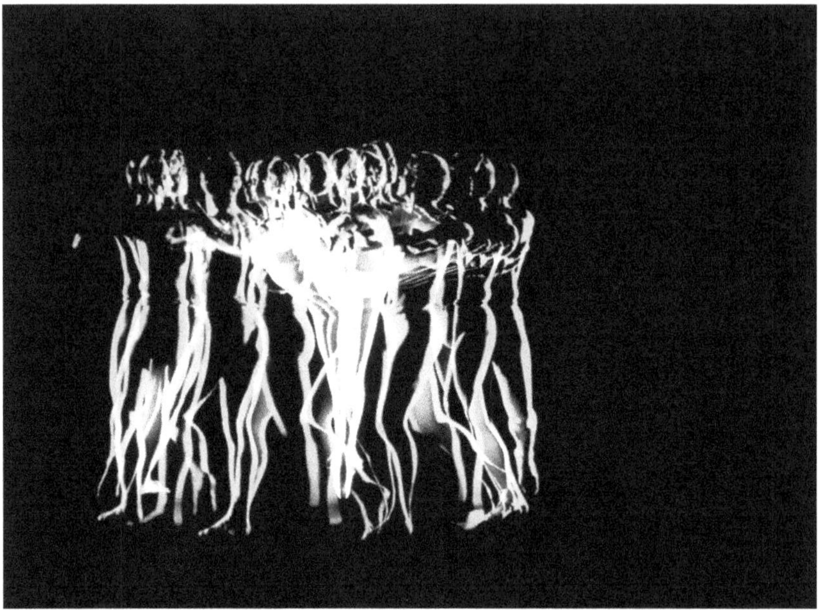

Figure 2.3
Norman McLaren, *Pas de Deux*, 1968.

here that McLaren's sensibility was proto-computational. The phenomenological time interval Manning identifies as inducing a sense of poetic grace is *also* one that is precisely programmable. In Thain's account, an experience of "heterochronicity," or a "multiplication of competing temporalities," is made up of a precisely calculated "combination of speeds and slownesses," the holding together of which is the mark of McLaren's ability "to directly animate emergence as the potential for the novelty of change itself."[60]

Birdlings 1967, Unreleased Computer Test with Evelyn Lambart and René Pardo

Kar Liang organized a series of computer demonstrations and seminars at the Film Board between 1967 and 1969. With invited speakers from the United States and borrowed computers, Liang staged a glimpse into the creative possibilities of computing for image production—not just for automated control over camera motion on the animation stand. During one of these computer demo days, held as part of the Expo 67 cultural program, Liang invited McLaren to experiment with the Digital Equipment

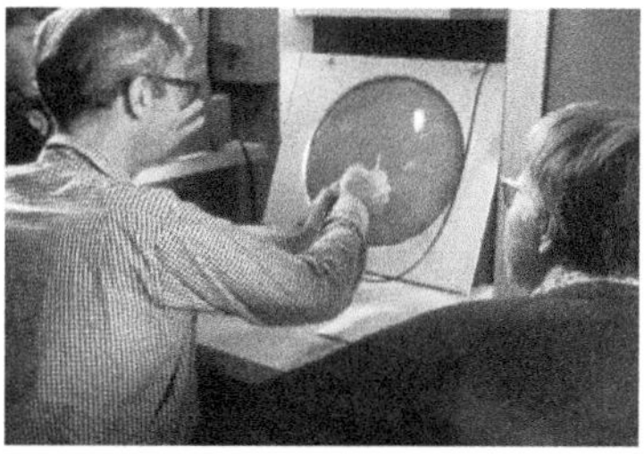

Figure 2.4
Norman McLaren at interactive display for test film *Birdlings*.

Corporation PDP-8 minicomputer he had obtained from the Data Division of the Department of Forestry. McLaren looked at the devices, then ordered everyone but Lambart, Liang, and René Pardo to leave the room. This turned into a weekend project, handled from a technical angle mainly by Pardo, a computer engineer interning at the NFB as Liang's assistant. Pardo recalls that McLaren was then working on the storyboards for *Pas de Deux*, so it was natural that he had the idea of multi-imaging in mind. "He wanted to control and visualize the glow that follows an object . . . but you couldn't control the phosphor. So he could not exercise complete control on the film."[61] The figures of McLaren's film resemble the birds from his 1958 film *Merle/Blackbird*. *Birdlings* dramatizes, in miniature, the scenario of *Pas de Deux*, starting with a soloist and ending with a duet. Leslie Mezei, a critic of computer art and University of Toronto computer science professor at the time, wrote that McLaren found the use of the PDP-8 computer, as responsive as it was for machines of the time, rather primitive. Still, Mezei thought, even though he manipulated just a few simple figures, "there is a great amount of humanity and charm in the film."[62] It appears likely that interactive computer experimentation in the *Birdlings* test film had a formative impact on *Pas de Deux*, one of the acknowledged masterpieces in the history of animation.

Proto-computationality and Artistic Research at the NFB

The concept of proto-computationality has come into focus in this chapter by a series of steps: first through the eyes of computer engineers captivated by the structural clarity and poetic beauty of McLaren's work, especially but not only seen in the films of the 1960s; next through the analysis of critics, outlining a more formal breakdown of the concept involving notationality in a quasi-musical sense, direct nonphotographic manipulation of materials, and algorithmic compositional design. Beyond the films selected to illustrate computationality, almost all well known in the film canon, the only test film McLaren made using a computer was described to show its similarity to one of his enduring masterworks.

Pierre Hébert wrote of his own creative process as an allegory of attunement between his tools and artistic signification,[63] an idea that might be more formally expressed as a techno-aesthetic equilibrium. Neither the content nor the technique predominates in this balancing act, so well demonstrated in myriad works by McLaren and his collaborators. Moving on from a narrow "proto-computational" aspect of McLaren's protean film practice—one that had demonstrably consequential impact on the computer technologists who were inspired by it—this account can now consider more broadly how a particular combination of human skills and cognition was nurtured within the institutional home of the NFB, in contrast to the way production tends to occur in industrial cinema based on a complex hierarchical division of labor.

McLaren's technological ethos was that of an artistic bricoleur and may be compared usefully with a contrasting approach to art and technology, that of the *rational artist*. The rational artist, according to A. C. Crombie, writing about Renaissance artist-engineers such as Brunelleschi, was a "virtuoso," not a bricoleur, exhibiting a mastery not in tinkering with "what comes in handy," but rather in a commanding control over materials to produce results based on an antecedent "rational anticipation of effects."[64] In contrast to the figure of McLaren as artistic bricoleur, it would be useful to consider briefly the early career of Edwin Catmull, the computer scientist who pioneered many of the foundational techniques of computer graphics and animation in the 1960s and 1970s, and went on to cofound and lead Pixar Animation Studios.

Catmull always knew he wanted to apply computing to the reengineering of the classic Disney-style feature film.[65] In his doctoral thesis work at the University of Utah, he created methods for the production of movies, aiming from the beginning at representing figures in three dimensions. The work progressed by digitizing real-world objects, then adding controls to program their motion. Catmull first made his mark by producing a test film of a three-dimensional animated hand, which garnered recognition—much as McLaren's *Birdlings* had—after being exhibited in the professional conferences of the computer community. As a demo film, Catmull's 1973 *Computer Animated Hand* made a stir because nothing like it had been seen before. Catmull's own scanned hand is first seen as a net of polygons, then opens and closes, then is shown smoothed-in with gray scale shading. The hand poses to point its index finger at the observer, beckoning her to approach. Next the beckoning gesture leads the spectator *inside* the hand itself, a fly-through of cavernous interior space, into and out of the fingers, that demonstrates a novel possibility of using computers to visualize not just the unseen but the unseeable.[66] The demo went on to play a cameo role in the Hollywood science fiction thriller *Futureworld* (1976). As noted by Andrew Utterson, Catmull's test film would become "the starting point and inspiration for countless CGI representations of the human form and other objects, now limited less by technology than by the imagination."[67] Whereas McLaren's *Birdlings* impressed its viewers by the whimsical charm of its motion dynamics, using the simplest of abstracted forms, Catmull's film demonstrated the initial feasibility of a research program aiming toward virtual realism, combined with the hierarchical stratification of work flow whose virtuosity is well illustrated by the hundreds of contributing artists and technicians seen on the credits of any Pixar feature today. Catmull, the rational artist, clearly anticipated the effects that would be incubated for twenty years before reaching fruition—first at the New York Institute of Technology, then at Lucasfilm, and then when Steve Jobs helped the core group of scientists (none of whom were filmmakers) establish Pixar.[68]

The concept of bricolage has been widely theorized and discussed since it entered the realm of human sciences through the work of Claude Lévi-Strauss. For the French anthropologist, the bricoleur was set up as an inverse foil to the scientist: where the bricoleur uses "events" to produce "structures," the scientist begins with structures and uses them to create events. The bricoleur uses "just what is at hand for a particular project."[69] This account of bricolage has been extended to the study of computer programmers. For instance, Sherry Turkle and Seymour Papert took pains to identify a bricoleur style of programming, more akin to the work of an artist than a scientist:

> The bricoleur resembles the painter who stands back between brushstrokes, looks at the canvas, and only after this contemplation, decides what to do next. Bricoleurs use a mastery of associations and interactions. For planners, mistakes are

> missteps; bricoleurs use a navigation of midcourse corrections. For planners, a program is an instrument for premeditated control; bricoleurs have goals but set out to realize them in the spirit of a collaborative venture with the machine. For planners, getting a program to work is like "saying one's piece"; for bricoleurs, it is more like a conversation than a monologue.[70]

What these authors call "soft mastery," however, should not be seen as a stage, but as a creative way of organizing work that *also* is applicable to science itself. Bruno Latour is famously incisive in refusing the modern "great divides"; indeed, there is no such divide at all between science and nonscience, but rather a "lot of small, uncertain, unexpected divides."[71] It is only a prejudice to think otherwise, according to Latour, and as soon as the prejudice loses hold, "cognitive abilities jump in all directions: sorcerers become Popperian falsificationists; scientists become naïve believers; engineers become standard 'bricoleurs'; as for the tinkerers, they may seem quite rational."[72]

In the case of Norman McLaren, the tinkering can indeed be seen as decidedly rational, as has been demonstrated throughout this chapter. Experimental art has been imbued with what critic Renato Poggioli called "technicity" and "scientificism" since the historical avant-gardes of the early twentieth century,[73] and it is sometimes observed that such experimentalism—enjoining creative liberty with technoscientific play and research—can yield breakthroughs of broader applicability than that of the individual artwork. McLaren's creative process offers another, more focused kind of object lesson in artistic innovation. The salient elements are: sustained stable support for a disciplined ongoing research program; iterative test-production cycling between artistic research and final artwork; circulation of results in technical reports; and an efficient permanent mechanism, in the form of a state film distribution system, for the worldwide circulation of finished artistic productions, as well as archiving of all the steps of the production process. That this was an artist of the *"arrière-garde"*—noncontestational and antitransgressive—may just be an incidental matter. It is still important to note that McLaren sought not to challenge the foundations of the art of animation, but to penetrate to its very essence in pure movement.

McLaren's exquisitely balanced mutuality in techno-aesthetic terms occurred too in the NFB's unique sociotechnical work culture: deeply collaborative and experimental, both artistically and technically, and organizationally stable, particularly over the years in question.[74] Besides animation research, the Film Board innovated continuously in documentary film techniques, film sound recording, and computer-controlled camera automation. For Expo 67, a team of animators and documentarists conceived the multiscreen Labyrinth pavilion, whose innovations would be carried forward to establish the IMAX film format. Gerry Graham, who served as the NFB's chief of technical research and was Kar Liang's first boss, documented these developments fully. Besides offering a historical account of this period, Graham also articulated the policy guidelines of the Research department, which are worth recalling today in considering the organization of creative and collaborative work in hybrid studio laboratory environments.[75] The mission of technical research at the NFB was "to gain new scientific knowledge with the primary purpose of applying such knowledge to solution of practical or technical problems":

1) Experimental development, in support of NFB production or distribution mandates;
2) Scientific data collection, attending national and international conferences, monitoring of state of the art;
3) Scientific information publishing project results in technical journals and seminars;
4) Testing and standardization;
5) Feasibility studies, special studies for NFB, other government departments, and private industry;
6) Scholarship programs—summer student program, international specialists.[76]

The first of these objectives, "experimental development," was ideally suited to the kind of systematic artistic experiment exemplified by McLaren, Lambart, and the rest of the artistic staff that comprised the animation unit at the NFB over this period. Engineer Kar Liang, who was electrified by McLaren's work at the common lunch counter the NFB shared with the National Research Council, would come to provide one of the crucial linkages needed to establish a formal collaborative relationship between the two federal agencies, a linkage between the research component of a cultural institution and the experimental art component of a research institution. The story of this collaboration, and its consequent spinoffs in the birth of Canada's computer animation industry, is the subject of the next chapter.

3 *Hunger/La Faim:* When the American Computer Had an Eye, and the Canadian Computer a Hand

This chapter examines the introduction of software as a creative medium for animation production at the National Film Board and the National Research Council of Canada during the 1960s and 1970s. While the early digital arts have been investigated broadly, fine-grained historical studies focused on the *transition* from analog to digital methods in animation are relatively few.[1] Animation—particularly fine art animation—provides a rich domain through which to study how creative practices were first modeled and computerized, as the predigital practices were both highly diverse and strictly conventionalized and modes of production were both individual and collective. As noted in the previous chapter, the hierarchical division of labor for cel animation, developed in the 1930s for the production of the classic Disney studio full animation feature films, beginning with *Snow White,* was paralleled by a strong tradition of independent experimental animators. The auteur tradition in animation was particularly strong in the period from 1950 to 1975, often in opposition to the elitism and hierarchical specialization of the industrial system, where there is one producer-director but hundreds of subordinate artists and technicians organized in a stratified work flow, producing complex cultural products with greater or lesser assignments of "humdrum" and "creative" roles.[2]

The early Canadian computer animation work grew out of the work at the NFB, analyzed in terms of proto-computationality in the previous chapter, alongside a parallel, more general research program for creative applications using computers at the National Research Council in Ottawa. In comparison with pioneering computer animation work carried out in the United States, including such artists as John Whitney, Stan VanDerBeek, Lillian Schwartz, Ed Emshwiller, Charles Csuri, Tom DeFanti, and Dan Sandin, and supported institutionally by corporations (IBM and AT&T's Bell Labs) and universities (Illinois, NYIT, Ohio, Utah), the Canadian innovations grew out of an alignment between two government agencies with complementary interests in experimental uses of interactive computing. The trajectory of this work spanned two decades and is divisible into two broad phases: (1) research system design and artistic production (1968–1975) and (2) technology transfer, deployment, and diffusion (1976–1988). During the first phase, the NFB and NRC collaborators came together around several systems for creative applications in the arts (the second one, beside animation, was for computer music, which is described in chapter 4), and produced a polished film for international distribution, *Hunger/La Faim* which was recognized with prizes at the Cannes Film Festival and an Academy Award Nomination in 1974—the first such recognition for a wholly computer-produced film.

In the second phase, the 2D software used for *Hunger* was handed over to a company charged with updating and adapting it for regular use as a production system at the NFB. The needs of the intended users in this technology transfer were incompletely understood and the system was delivered in an unworkable state. After debugging, and only occasional uses in production, the system was then replaced by turnkey 3D software packages running on newly available workstations in the mid-1980s. The human capital developed in this phase helped set the stage for a start-up business culture in Canada that fostered a set of internationally successful 3D animation firms selling applications designed expressly to serve the needs of nonprogramming artist end-users. Within the NFB, the arrival of a market for animation production software in the 1980s was seen as a trigger to replace a unique homegrown Canadian system with industry-standard hardware and software. The ensuing controversies inside the NFB turned on whether to accept what was seen by some as a suspect ideology of digital progress associated closely with the supposed "advance" from two- to three-dimensional computer graphics. Thus the Canadian case allows a privileged vantage point on the history of analog-digital transitions in a specific creative practice, covering a trajectory that spans all the way from early "mainframe experimentalism"[3] to the widespread diffusion of user-friendly packages for production by nonprogramming artists, and which is sensitive to a variety of critical views and retrospective assessments of the aesthetic gains (and losses) that occurred at various stages along the way.

Artists as Lead Users of Early Computer Animation Systems

The base technologies for interactive computer graphics were largely developed in US military research programs, often closely aligned with key universities like MIT, and supported by the Pentagon's aggressive funding of fundamental information processing research.[4] By the mid-1960s, development of civilian applications was under way as well, notably in aviation, architecture, and scientific communication. Broadly speaking, two design approaches to computer animation were initially pursued: picture-driven and language-based. The latter specified visual images and their continuity using traditional textual computer programming languages; they depended on the ability to describe visual phenomena mathematically. Picture-driven approaches aimed to assist aspects of the hand-crafted art of animation, permitting the artist to draw the cels serving as keyframes, using the computer to coordinate the images and calculate the transitions between the keyframes.[5]

Beginning in the mid-1960s, researchers at the National Research Council of Canada, the principal basic research laboratory of the federal government, began to investigate the potential for using computers in filmmaking and music composition.[6] In both cases, the Canadian investigators were scientific and technical followers, not leaders. In comparison with their American counterparts, they had very restricted budgets for equipment and personnel, and were therefore acutely aware of the need to select their area of intervention with caution. They began their research, as Kar Liang had done at Northern Electric, by systematically studying everything the Americans had done to date, with the aim of finding unique research problems appropriate to their local work culture, and especially to their restricted means. Though the notable success of this work came from the ways the NRC and NFB found to cooperate with one another, the programs began separately and grew up within research cultures sympathetic to techno-aesthetic experimentation.

In the case of the NRC, the Radio and Electrical Engineering department had been the location of Hugh Le Caine's groundbreaking research in electronic music since the early 1950s; this work is described in greater detail in the following chapter, in connection with the importance of concepts derived from Le Caine's "playable" electronic music instruments within the emerging field of human-computer interaction. The NFB Technical Research

Program had an ongoing responsibility to monitor the global state of the art in cinema and audiovisual technology on behalf of the Canadian motion picture industry as well as the government. During the mid-1960s, when creative investigations of computing began, NFB technical innovation was internationally recognized in recording, projection, and editing methods. Both its technical staff and the animation and documentary units employed resourceful producers who were themselves individually committed to technical experimentation as a regular aspect of production. The range of problems they needed to solve included low-cost production for documentary and scientific filmmaking (often of technical/mathematical subjects) as well as high-profile international expositions—showcases of state-of-the-art cultural technologies in world's fairs like Expo 67.

The NRC researchers, Ken Pulfer, Marceli Wein, and Nestor Burtnyk, had previously worked on radar and rocket instrumentation, and did not consider themselves even to be "computer people" in 1965, when they decided that making computers usable by "nonliterate" people was an interesting problem. (By this they meant, scientists and engineers who would use experimental computational methods in their research, not as routine clients of data-processing facilities.) From the start, they had little interest in the batch-based methods of mainstream computation; indeed, their interest was not in number crunching per se, and certainly not in the theoretical subject matter of the newly forming discipline of computer science, but in "problem solving in a general sense."[7] Coming from a background of tinkering with rocket hardware, what they wanted was speed, responsiveness, and in particular real-time graphical display. They made the unorthodox choice of acquiring a general-purpose industrial control computer, then designing and fabricating the input-output devices necessary to support their two chosen creative applications: animated movies and music composition.

The nature of the work process they wanted to support was close to what Donald Schön calls the "reflexive conversation" of any skilled practitioner with her materials.[8] On the supposition that creativity in applied and pure sciences is the same as in art, Pulfer elaborated, in a contribution to the new journal *Man-Machine Studies*, that "the process involves trial and error, with the creator modifying the mental image of his creation as it takes place. He interacts with his creative medium . . . in a conversational way, learning the 'language' in which he can express himself as he goes along."[9] Pulfer and his colleagues' primary interest was not mainly to develop applications in computer graphics or music as objectives in themselves, but rather "to find out more about the interaction problems in a creative environment." He recalls their aims as discovering

> how to help physicists and other scientists who were starting to use computers. Some had that kind of mind to program, others didn't. So this wasn't a surprising area for a national laboratory. Going from the needs of scientists to animators and musicians was fine, we were trying to generalize the problem. The scientists were being supported at the regular computation center; but they weren't being helped the way we were talking about.[10]

Reviewing what had been achieved to date in the US labs, they determined early on that there was little point in using the systems themselves to create proof-of-concept demonstrations. Indeed, they found the artistic quality of much of the American engineer-programmed artistic efforts lamentably pretentious and embarrassing. Their own latent creativity was, it appears, never idealized, nor the computer taken as lowering conventional assumptions of high barriers to creative originality. The more serious challenge for them was instead to support "real users"—that is, professional composers or animators, but in a work process entailing incremental, conversational, and direct hands-on improvement. Pulfer wrote at the time:

> Up to this point, it has been assumed that the best possible way to design the computer would be to make it transparent. That is to make it look to the user as though it were not even present, so whatever idea occurred to him, it could be rapidly formed into a final creation. This is not necessarily true.[11]

Instead, Pulfer insisted, constraints are crucial to the creative process, giving examples such as conventions for perspective drawing in architecture, or scales, chords, and notational schemes in music. By computationally modeling such conventions, the user is given a more meaningful starting point than the blank slate of total generality.

> Most computer languages now available . . . are unsatisfactory either because they are mathematically oriented, or because they result in cumbersome and slow programs. As a result we are usually left with the situation where an artist-programmer team is formed, the artist uses the system without having intimate control over the functions of the blocks he uses, and the programmer builds blocks without fully appreciating the needs of the artists.[12]

Pulfer and his team therefore undertook to develop a system in which

> at no time [would it be] necessary for the user to learn how to program the computer, or in fact even to know how to operate it other than through making some choices from names presented to him on the screen . . . he can proceed to learn the "language" by trial and error.[13]

Crucial to the implementation of this design was the just-published research of the first graphical user interface demonstrated in 1968 by Douglas Engelbart of the Stanford Research Institute, as a system for "augmenting the human intellect."[14] It is now well known that the researchers in Engelbart's group carried the main innovations forward to Xerox PARC where they became important aspects of what would become the personal computer interface.[15] It is less well known that the NRC animation and music researchers were actually the first outside Stanford to reimplement the mouse, as well as the chording keyset for the left hand, as initially conceived by Engelbart. By 1970, as Canadian musician Bill Buxton (now a principal scientist at Microsoft, and who is a key figure in the next chapter) recalled, the NRC's homegrown assembly was the functional equivalent of a "1984 Apple Macintosh," in use around the clock by professional composers and animators.[16]

In turn, the animation project came into sharp focus when Burtnyk heard a Disney studios animator in Los Angeles describe how the factory-style animation studios operated, with work divided between head animators and assistants who filled in the in-betweens.

> Understanding of the Disney studio was crucial here: the in-betweeners were not low-level assistants, but they weren't providing the creative component either: rather they were restricted in what they can do, their contribution is very well defined, precisely constrained, to just do the job as required. . . . The nice thing was, it was not inventing a new way to do things, we were just going to mimic the conventional process with which all the film animators were familiar.[17]

By 1970, the NRC had implemented for the first time a complete user-oriented system to emulate conventional cel animation, where the components are drawn on celluloid and combined into sandwich layers. This eminently practical orientation would use the computer to simulate the drawing table and mimic the traditional tools used for drawing, inking, and in-betweening of traditional cel animation. The system was immediately put to use for commercial title sequences and was ready, in the researcher's minds, to take on more challenging creative projects.

The Production of *Hunger/La faim*

As seen in chapter 2, the NFB maintained a strong technical research department to monitor the global development of motion picture technology, and this group too had a well-established tradition of technical innovation. Kar Liang joined the Film Board in 1966 and applied his previous methodology as electrical engineer in the telecommunications industry, on the application of the computer to digital signal switching, to a disciplined benchmarking approach to the analysis of the computer as a tool for motion pictures. This quickly turned into an intensive learning program in which the NFB received visits from, and in most cases pursued in-depth dialogues with, all of the key US players; it also conducted tests using borrowed equipment, including the weekend blitz that produced McLaren's *Birdlings*.

As discussed in the previous chapter, McLaren taught that the animation setup should be malleable to the individual's creative gesture, permitting improvisation like a musical instrument; but most important was that the particular material properties of the setup, their limitations "when brought in touch with the theme," should "be the growing point for visual ideas."[18] This was at the heart of McLaren's artisanal lesson to NFB animators, that each new film be made in a small workshop of at most a few persons, with its own unique technical approach.

René Jodoin, the head of the French animation unit, was quick to understand the way a computer technique for automatic in-betweening could enhance the already very heterogeneous range of techniques and styles pursued in his unit. Rather than considering it as a generic solution applicable across the board to all problems of production, he considered it as he would any other resource that would be applied to a specific creative proposal. Within the NFB, the aesthetic interest in computing had so far been articulated in terms either of increased precision in the control of the camera on the animation stand (growing out of the needs of science education), or computational modeling of image vocabularies using algorithmic logic influenced by avant-garde musical practices. Neither of these existing interests were relevant to the properties of the NRC system, which took as its ideal user the traditional character animator accustomed to delegating the in-betweening task to highly skilled assistants.[19]

As it happened, Jodoin had just received an unusual script proposal from a freelance animator then living in France, Peter Foldes, which called for the extensive use of metamorphosis in the context of full hand-drawn character animation. This script, entitled *Hunger*, would have required a fairly large production team including many assistants. This would have been more costly than normal, and using this style in any case went against the grain of the prevalent production philosophy in the French animation unit of Film Board (cel animation was more commonly practiced in the English unit). But to Jodoin's mind, Foldes's script, demanding extensive use of metamorphosis between images, might just be a good fit for the already-demonstrated capabilities of a system designed for automated interpolation.

Jodoin next commissioned a staff animator in his unit, Pierre Moretti, to document the system in detail, and to work with the NRC designers to make a series of improvements prior even to beginning any production. These exchanges of NFB official memoranda circulated throughout the studio and with the NRC. Soon a pattern of mutual accommodation developed between the NRC researchers and the NFB artists in which the system would be regarded not as a fixed tool, but as a resource to be modified in response to particular developing script requirements. As Jodoin put it in retrospect:

> I think there's a distinction between using a tool, and something able to give results beyond traditional means. Because of the involvement of scientific types, Nestor and Marceli, the artists were able to work together [with them] in such an intimate way. . . . There was an understanding on their part, to satisfy Foldes and myself. . . . It was so beautiful, because it was so hard to distinguish everyone's contribution in there.[20]

In this pre-production phase, it was evidently important in the egalitarian work culture prevalent at the NFB to describe the new apparatus in terms that would be intelligible to the entire group, whatever their individual technical preferences. A demonstration for Foldes was arranged, and in two afternoon sessions a group of sequences was completed by Wein, Burtnyk, and himself that could be edited into a compelling demo film—*Metadata*. Moving quickly to share the results of this work with the international animation community, Jodoin approved the entry of *Metadata* for the main international competitive festival for animation at Annecy, France. In spite of taking just a short burst of effort to produce, *Metadata* garnered significant attention and reviews in the summer of 1971. This encouraging response led Jodoin to insist that the next step should be to make a "normal" NFB film, that is, one that went through the usual process of peer-based approval—an involved multistep bureaucratic process.

Apparently aware of the special risks in displacing production from the NFB's Montréal studios to a scientific laboratory in Ottawa, a set of explicit criteria was prepared to guide and justify the selection of a film proposal suitable for a computer system. Jodoin assigned this exacting task to Moretti, who produced a highly analytical, nuanced memorandum that was circulated to the whole unit. A candidate computer film should be one based on a "valid cinematographic concept"—i.e., not merely a catalog of technical effects—with a scenario that would uniquely "lend itself to solution by means of computer-controlled animation." It must also present an aspect of research concerned with advancing the generic possibilities of the system, while still being suited to the system's quite limited capacities. It should be chosen to push the medium enough to yield "generalizable" results applicable beyond the single instance; and finally, it must be presented by a "creative talent with the background to make a coherent and structured piece."[21] In the egalitarian work culture of the NFB, only after Moretti's memo was vetted did the full production approval for *Hunger* go forward.

Peter Foldes had attended one of the international meetings on new film technologies organized at the NFB in Montréal and was already enthusiastic to experiment with electronic techniques so as to animate his hand-drawn figures. Unlike the early artist-users in the United States, who were either modernists in their rejection of figurative drawing or engineers who wanted to computerize animation precisely because they could not draw, Foldes's artistic forte was his superb draftsmanship. Hungarian by birth, and an émigré with studios in London and Paris, Foldes was noted for making films with politically charged subjects. *Hunger* is a searing denunciation of human inequality and economic exploitation, depicting a runaway consumer society eventually devoured in a revolutionary uprising by the hungry and oppressed. The film's protagonist is the very figure of civilized gluttony, robotically and inexorably driven to gobble up everything before him. The uncanny mechanism of the smooth interpolation techniques is vital to producing the deeply unsettling mood of this dark vision. The storyboard treatment (made before computation was proposed) used metamorphosis fittingly to dramatize the self-destructive fate of civilized man. In this way, *Hunger* was consonant with the public education tradition of the NFB, epitomizing a late-1960s mood of sociocultural crisis and protest, and was sharply distinguished stylistically as well from the preponderantly cute child-oriented mainstream Hollywood animation product. In its socially charged subject matter, it was also markedly different from the mainly abstract efforts of John Whitney, the Bell Labs-supported artists Stan VanDerBeek and Lillian Schwartz, or the various other mainly abstraction-driven projects undertaken contemporaneously at the University of Illinois, Chicago campus, and at Ohio State University. Pierre Hébert, again an astute observer of animation aesthetics within the NFB's unique culture, noted:

> It was the genius of René Jodoin that he could see from the possibilities of the NRC system, that it was precisely the style and the visual research of Peter Foldes which could directly and immediately make use of it.[22]

The artist spent a few weeks at a time working with the system in Ottawa; in the intervening periods, improvements were made based on what had been learned in production. When *Hunger* was released in 1974, it was recognized immediately as an artistically convincing character animation, garnering an Oscar nomination and a grand prize at Cannes. André Martin praised *Hunger* as the first convincing character animation using computer methods, calling its depiction of human figures credible and convincing.[23]

The accomplishment of *Hunger* in matching an artist's vision to the still very intractable computer of the day can be interpreted in a number of ways. For one, the technique of linear key-frame interpolation was still far too primitive and mechanical to be used for the "anthropomorphic" style of the big-budget feature animation studios in Hollywood. (Recall too, from the previous chapter, how according to Evelyn Lambart, NFB animators were openly contemptuous of Disney, and also had expressed solidarity over the years with Disney studio labor organizers.) While it promised to save costs by automating the intensive human labor of the artist drawing the intermediate frames, given its technical awkwardness, it could only be put to creative use by an artist willing to shape her vision to its mechanical constraints—a requirement for which Foldes precisely fit the bill. Indeed, this "machinic" interpolation, which in conventional contexts would have been a fatal defect, gave the film its expressive signature.

Film scholar Mark Wolf comments on the peculiar expressive qualities of the linear method of interpolation in *Hunger*, nothing that "during transformations in Foldes's films, lines move independently of one another . . . one object could metamorphose into any other object, regardless of how different they initially appeared from each other. In this sense, Foldes's work anticipated *feature-based morphing* in which correspondences are chosen by the artist, allowing objects without any inherent similarities to be morphed from one to the other."[24] The contour edges of figures and objects are subject to continuous, often surrealistic change, not just treated by displacement as rigid entities in space. Their very ontologies become unstable: a face becomes a flower becomes a telephone, in exactly the sense of metamorphosis as the essence of animation in Emile Cohl's 1908 *Fantasmagorie* (a work often singled out for praise by Norman McLaren).[25]

Furthermore, the impact of the film proved to be far-reaching. It demonstrated that convincing artistic films *could* be produced by computer, at a time when Hollywood was only using computers, if at all, for title sequences or rudimentary special effects. As well, it had a major influence in the technical community, attracting young people, especially in Canada, to the field of computer engineering or filmmaking precisely to further the possibilities of artistic animation. Daniel Langlois, Doris Kochanek, Catherine Richards, and many more experts outside the Canadian animation

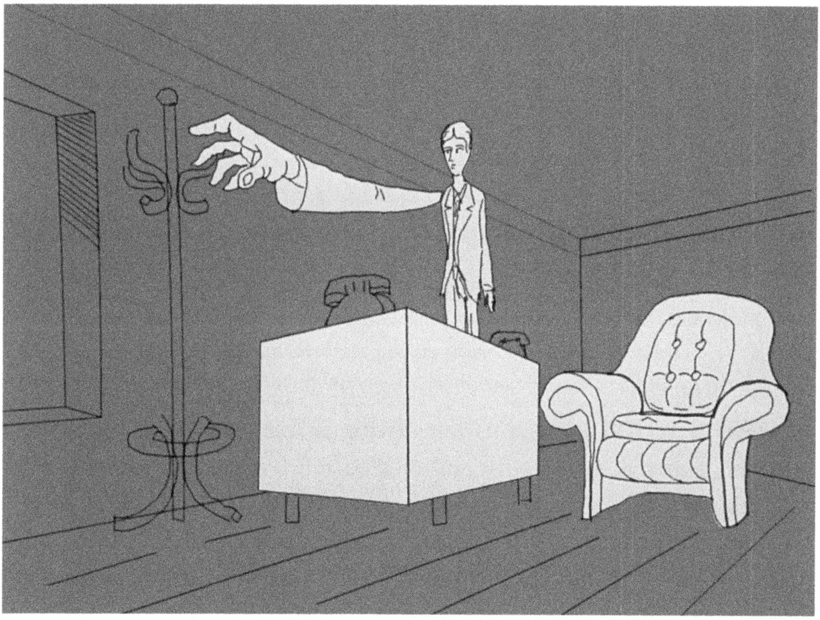

Figure 3.1
Peter Foldes *Hunger*, 1974.

community, such as the attendees of the SIGGRAPH computer graphics conferences, were turned on to creative animation potentialities by *Hunger*. Ed Catmull, then at the New York Institute of Technology, took this film as a cornerstone example in his landmark 1978 paper on the problems of computer-assisted animation; pointedly, he acknowledged it as an aesthetic success, but derived from it a cautionary lesson, namely that the problems of 2D are much more challenging than 3D animation, since 2D itself is a projection of 3D characters as visualized by the animator; hence "information is lost."[26] For Catmull, this meant that improving on systems like that used for *Hunger* would be like solving difficult problems in artificial intelligence; better, he said, would be to focus initially on the more puppetry-like problems of 3D object manipulation. Two decades later, Catmull would be instrumental in supporting the Academy Awards for Technical Achievement garnered by Burtnyk and Wein in 1997.[27]

In this early episode of productive collaboration between two participating organizations, both were small, underresourced, and unable to make further progress without the contributions of the other. None of the researchers identified strongly with (nor necessarily even knew) the way things "ought to be done" in computing. From the outset, both organizations had something of a hybrid character—the NFB, a cultural organization with a strong technical research group, skilled at absorbing and repurposing new techniques; the NRC, a government research institute with an intellectual work culture friendly to artistic practice.

Foldes, whom we can consider the lead user of the NRC system, realized how unusual was his opportunity, when he commented during the production of *Hunger*:

> One could say the *American computer has an eye, and the Canadian computer, a hand.*[28] The Americans have commercial pressures, a concern for profitability. *The Canadians at the NRC are much more disinterested, and finance pure research.*[29]

After they had made this convincing artistic and technical demonstration, Burtnyk and Wein continued to refine the system, addressing a specific control problem that had only arisen through the production of *Hunger* and the extensive discussion the work had provoked in the technical community. Professional character animators found the smoothness of the system's metamorphosis problematic, as it treated shapes as global but did not allow small nuanced selection of differential rates of change in subcomponents. To address this deficiency, they introduced for the first time the construct of an intermediate "skeleton" with a given number of control points which dragged the outer skin closest to them as they were modified.[30] Thus the animator was able to introduce refined variations in the rate of change within parts of an aggregate, without needing to formalize each of them. The 2D skeleton concept, like the mathematics of splines, allowed the artist to draw structurally in a new way, which greatly improved the refinement that each keyframe was given before the automatic in-betweening was carried out.

By then, their research was clearly legible in the professional terrain of the newly formed computer graphics research community that was founded during the years of production. These research results on skeleton drawing were duly presented to the ACM SIGGRAPH conference in 1976[31]—and even recognized as one of the seminal scientific papers in the field 25 years later. However, this much-improved version was never proven out through the rigors of creative production as the first system had been with *Hunger*. Burtnyk and Wein did not see themselves then as HCI specialists (though that is what they were, *avant la lettre*) because that field itself was in a predisciplinary formation. By producing this paper, Burtnyk and Wein had moved fully from general investigators seeking to understand intuitive techniques for human-computer interaction, to builders of advanced motion dynamics programming systems.[32]

Technology Transfer and Deployment in Production

At this point (1975–1976) the NRC discontinued its officially mandated research into artistic applications. The reasons are several and conflicting: management suggested they needed to look at industrial applications of graphics, such as robotics or space exploration; evidently, the notion that animation software would itself become an important industry sector for Canada was not even apparent to anyone at the time. It must be recalled that during the 1970s, a market for third-party "application software" was still nascent.[33] Facing rising economic pressures from the mid-1970s, the NRC had also become uncomfortable about its association with art production, and this discomfort was expressed in practical language as needing to spin off the system to the private sector having successfully made a convincing demonstration—the first in the world—of the viability of computer-controlled character animation. The pure research climate so praised by

Foldes was starting to become constricted—much as was happening over the same years at Bell Labs in New Jersey.[34]

Wein and Burtnyk moved ahead nonetheless, by arranging for a technology transfer of the prototype production system to the NFB on the assumption that the animators there could continue to work with it on their own once it was updated for faster hardware. The transaction, paid for jointly by the two agencies in a deal worth about a million dollars, was difficult and took several years just to negotiate, and another three years for the work to be completed by an independent contractor.

The NFB too found itself unable to understand what to do next with the use of the computer in animation, despite—or perhaps because of—the unexpected success it had enjoyed with *Hunger*. As Mintzberg pointed out in his study of NFB administration over several decades, the relative consistency of the late 1960s period, during which a broad consensus emerged to privilege a strategy of "experimentation," was followed by a more confused and disordered time of internal dissension.[35] Factions now competed to define the future validity of computing in filmmaking at large. Three distinct competing stances can be discerned: some of the creators took a limited functionalist position, agreeing to the eventual use of automation for administrative purposes only; the technical research group, lacking the initial vision given it by Kar Liang, argued for the integration of computing in production solely for purposes of image storage and the automation of editing; finally, somewhat marginalized by these other positions, Jodoin and the experimental animators argued in favor of continued creative use of computing at the Film Board by investing in the technology transfer arrangement proposed to it by the NRC.

Ultimately Jodoin prevailed, but only after a delay of about two years. One of the difficulties of this transfer was the choice of the firm contracted to do it, Digital Methods Limited, which had prior background primarily in military contracting but very restricted access to the future users at the NFB. Digital Methods was apparently never actually convinced that the system would ever be placed in the hands of the naïve user who was not also a "computer person," and the program was *never even tested* before being delivered to the NFB in 1979. The idea of insisting on a direct involvement by the filmmaker throughout the design and development process (which had been so carefully thought through by Jodoin and Moretti and sustained with Foldes) never happened. Thus, from 1976 to 1979, no progress was made along the dimension of successive or progressive refinements in use. And these were the very years, in competitive terms, that major investments were being made in computer animation in the United States, particularly at the New York Institute of Technology, which had assembled a cutting-edge research program co-led by Ed Catmull and Alvy Ray Smith.

So it was that the new system delivered to the NFB was not even functional, much less fully documented for practical use by filmmakers. The French animation unit had no computer personnel capable of debugging it, so it sat unused for months. Burtnyk at the NRC, by then promoted to a management role, was frustrated at the minimal performance of the Digital Methods contractor, and later called the system a "botch" as it had paid no attention at all to its usability by filmmakers. Several months passed before the NFB unit head, Robert Forget, was able to sponsor intern engineering graduates from the University of Waterloo to catalog its errors, repair, document, and test it. One of them, Doris Kochanek, continued to refine the system, much as Burtnyk had done earlier, by providing ever finer facilities for control over differential rates of change at the level of subcomponents of images; this involved a mathematical generalization (splines), encapsulated with control parameters that an animator could manipulate numerically.[36] Two other interns with design training, computer novices but game to experiment, made title sequences, calibrated colors, and wrote manuals in English and French hoping this would attract studio filmmakers' interest. One in particular, Daniel Langlois, made a point of driving the system in ways that went far beyond what it had been designed to do. Although meant only for 2D, with a vector rather than raster display, he used the program to simulate shaded 3D objects, including interpolation—a painstaking and counterintuitive repurposing task requiring the production of huge numbers of keyframes.[37] It also gave Langlois his first experience in 3D modeling, though using a tool inherently unwieldy for this task, so that he was able to learn a great deal from it about what a 3D drawing interface using splines would need to do in order to pass the threshold of general usability. Two official NFB productions were completed during the five-year period of incubation. One was in the tradition of precise character animation initiated with *Hunger*, but without the evident dependence on metamorphosis which appeared to be a trick in the eyes of some of the NFB artists—too much of a "one-liner" effect, and therefore too easy just to pull off the shelf. The other film, made by Jodoin himself with Kochanek and

Langlois, was a geometric tour de force in color flicker, *Rectangles*, which also ignored the features of the system that made it possible to produce character animation: his interpolations were strictly mechanical, regular, and geometric, and in a stroboscopic flashing rhythm. *Rectangles*, in the tradition of Conrad's *The Flicker*, is a work of "neural" cinema—a cinema of the brain, as Deleuze put it in *Cinema 2*:[38] a hallucinatory masterwork fully worthy to set beside *Hunger* as an early landmark of computer animation coming out of the NFB.

Still, the unwieldiness of the system made it unattractive for expressive purposes to the regular NFB staff animators other than Jodoin; far from being widely appreciated as an important new general resource, to be reconfigured and recombined with the repertoire of other techniques in the studio, it remained isolated in the small circle of hardy early users like Langlois, plus Jodoin. Yet despite the ineffectiveness of the technology transfer process and the relatively small development resources available to it, the NFB system was still as advanced as, and in some respects ahead of, any available alternatives at the time. In his interview with me, Langlois attributed this to the drawing affordances and the skeletal motion dynamics that permitted visual molding of the changing structure of objects rather than doing so by mapping strokes or writing code.[39] This particular contribution was one of several that built up a mass of expertise and insight permitting Langlois to see what would be needed to advance beyond the partial solutions then available, and to be ready to innovate in software animation tools.

Meanwhile, by the late 1970s a widespread anti-technology backlash in the arts had swept aside the enthusiasms of the previous decade.[40] At the NFB, the disillusionment was notably fueled by revulsion against the stereotypical use of hard-surfaced, mechanically animated 3D trademarks and logos in commercial advertising: the rigid movement of lifeless objects appeared to some the very antithesis of the sort of expressive flexibility and improvisational immediacy they favored. This tendency in commercial production grew quickly in the early 1980s as computer generations changed and the new UNIX-based workstations, attached to high-quality raster displays, swept aside all the minicomputer-based systems of the DEC PDP class in studio production. By 1984, the first turnkey commercial 3D animation packages were available for these UNIX machines. For the flag carriers of the McLaren auteur tradition, the simple-minded naïveté of the early attempts at realistic scene simulation and "flying logo" graphics were so regressive aesthetically that the entire area of digital technology was seen as tainted. This was not a naïve technophobia, but a resistance by technically sophisticated artists to an unnerving reductionism they saw in the culture of the 3D world. Recall also, as noted in Catmull's important article, that 2D is really "harder" than 3D in many ways; the explicit third dimension takes away a subtle level of artistry: this meant "more" dimensionality but "less" freedom, as noted by Pierre Hébert.[41] Hébert himself took a radical about-face from his early formalist interest in computer graphics, as he turned to scratching his films directly in real time on celluloid and improvising alongside jazz musicians, particularly musicians in the Québec *musique actuelle* scene and American experimental musician Bob Ostertag.

By 1984, however, Forget had become less oriented toward sustaining the internally driven artisanal traditions of his internal staff, and more concerned with keeping up with the fast-developing advances in the international computer graphics community. In a controversial decision that divided the studio sharply, he mothballed the NRC-derived system, mainly on the grounds that to be compatible with the leading edge of computer graphics, the NFB would have to invest in the next wave of UNIX workstations and programs for 3D modeling and animation. Langlois and the other computer experts were keen to implement this costly upgrade, which was made possible by special funds only made available for producing a single 60-second 3D sequence for a challenging stereoscopic IMAX film in the 1986 world's fair in Vancouver. The new Montréal facility, formally named the Centre Animatique, was opposed by the more artistically radical veteran animators. Jodoin, for instance, felt that 3D was less interesting artistically: too easy, he thought, as it tended to regress to puppetry and therefore the conventions of narrative cinema: he wanted instead to experiment more with unorthodox approaches to interpolation, still within 2D.[42] Hébert too, a keen reader of Walter Benjamin, saw the forward march into 3D as an unwelcome insinuation of an externally measured and false notion of progress, a forced conformity with mainstream entertainment values that he saw as responsible for the dubious rise of a culture of simulation. This also was a betrayal of what he understood to be the core value of homegrown artist-driven technical research.

Quite the opposite view was taken by Langlois and the other internal "computer people" at the Film Board, all graduates of the University of

Waterloo's elite computer science program. Even if Forget was himself somewhat confused about the reasoning, making self-contradictory statements in print, the group of young computer experts was able with his sponsorship to expand into a full-fledged department of computer professionals, and thus to join into and even take a leadership role within the mainstream trajectory of the industry. Forget had tried to justify the large expense and investment in staff on economic grounds, arguing it would make possible the production of internationally exportable TV-grade (i.e., low resolution) cartoon series. This corporate-directed rationale was, needless to say, of little interest to the old-guard auteur animators. Nor was it believable even to the computer people, who knew that 3D computer production, at the level with which John Lasseter and his colleagues at LucasFilm were starting to show promising results, was still far from cost-effective except for limited use in special effects—the very antithesis of the NFB animation studio's artisanal bricoleur ethos. Kochanek, the lead technical staff member, gamely promoted the Centre Animatique as a service for "research and development, integrated into a production environment . . . to be on the leading edge technically, but to do it in a production context, not as a purely R and D type of facility. So we would take on a couple of projects, and through those projects, push the state of the technology."[43]

The Centre Animatique gave the NFB the equipment and personnel base needed to become a capable player in the rapidly forming professional field of high-end 3D animation. The application software, acquired off the shelf from the US startup software company Wavefront, gave Langlois an opportunity to be the among the earliest adopters of a high-end turnkey system. The Wavefront system was one still designed for use by skilled programmers, and Langlois could see after intensive experience with it that integrating its various modules into a seamless "Macintosh-style" graphical user interface would make it accessible for the first time to artist-animators unskilled in traditional programming languages. Neither a skilled character animator nor a fluent programmer himself, Langlois combined the unique insights he acquired in the two phases of the NFB animation systems with a detailed knowledge of the difficulties animators would face manipulating complex organic 3D shapes. He gained invaluable experience in these efforts by serving as art director in a team that produced one of the first plausible 3D character animations, *Tony de Peltrie*. This film was programmed using the University of Montréal's mainframe computers, primarily by Pierre Bergeron,

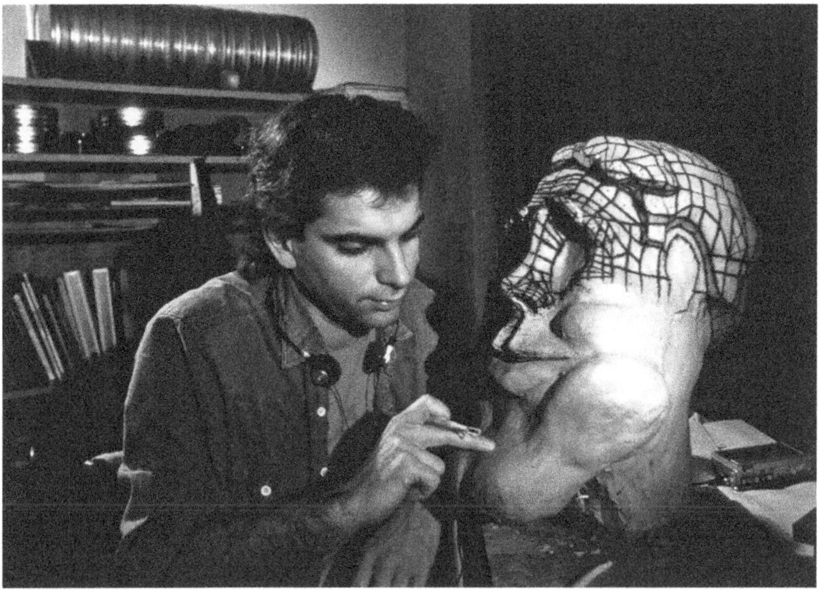

Figure 3.2
Daniel Langlois in production *Tony de Peltrie*, 1984. Courtesy of the Daniel Langlois Foundation.

and produced by Philippe Lachapelle. Depicting a sentimental piano player, *Tony* gave Langlois further inside knowledge of the difficulties of animating characters in 3D, particularly focusing on facial expression. The work first involved Langlois producing digitized versions of a large number of expressions; then these expressions were mapped on the caricatural digital face.[44] To the core group of leading-edge 3D researchers at what would become Pixar Animation Studios, this film was more than just technically interesting, as it provided a basic sense of "living and feeling" in the lead character.[45]

Langlois would go on to borrow the capital needed to found one of the most successful turnkey 3D-animation packages—Softimage—in 1986, just after he left the NFB. Though overlooked in the official histories of the corporation, Langlois's startup efforts were aided vitally by artist Char Davies, his then partner, who would go on to become Softimage's director of visual research. Her later achievements in the breakthrough works in virtual reality are discussed in chapter 7 of this book. Softimage was considered the first to provide what the industry calls integrated work flow for modeling and animation, allowing individuals with no engineering background, or firms without technical staff, to work independently.[46] It also offered inverse kinematics—a method of simplifying gestures from the end point of limbs—at a time when this was not available from other vendors.

Langlois attributes his exposure to the strong auteur tradition of the NFB as an important influence in insisting on a system that would permit the artist to be the "maître-d'oeuvre," yet the premium Softimage placed on ease of use would also distance the animator from an intimate closeness to the program's mechanisms in a way that was antithetical to that auteur tradition. It effectively responded to the commercial demand for adding movement to articulated quasi-organic characters—initially dinosaurs, insects, toys in the main—but in so doing, it closed off the black box of the program to tinkering or custom scripting entirely. In this way, following the account of Carolyn Kane writing about digital color, Langlois participated in the movement from open systems "close to the machine" to so-called transparent ones that are only democratic in an ironic sense, since these are systems catering to people who can make them work immediately, without knowing much about *how* they work.[47]

From the critical standpoint of Hébert, one of Forget's more philosophically inclined opponents in this regard, the entire episode was in this way a betrayal of the original vision of Jodoin, who had attempted to privilege direct access between producers and relatively simple means. Forget's ambition to place the NFB at the "leading edge" of digital development, competitive with the main US centers, including the ambition of developing new systems, was in retrospect unsustainable, and none of the films produced from that time had a success remotely comparable to that which *Hunger* had earned 15 years earlier. For Hébert, indeed, the adoption by the NFB of the dominant trajectory of 3D simulation indicated a more serious betrayal of the NFB's artisanal ethos, as it would ultimately accept the mystification of a certain ideology of the digital that attempted to validate reality only by simulating it. Hébert elaborated on his Baudrillardesque critique of the culture of simulation in a set of essays published in the late 1990s. With some bitterness, and also resignation at seeing the flourishing direct animation tradition of the past decades replaced by what he saw as skilled manipulators of powerful simulation engines, Hébert insisted on the need to disconnect what he called the "pensée technologique" of McLaren from any notion of linear technical progress and of instrumental rationality:

> The fact that digital techniques provide a vast range of possible choices constitutes [another] illusion, directly linked to the idea of immateriality and virtuality. It is said that it is now possible to explore thousands of alternatives to find the so-called "best solution" . . . I don't think that for every artistic problem, there even exists a "best solution." I think, on the contrary, that for every problem there always are many solutions that could be good or bad according to the network of coherence/meaning that associates this with other choices, a network which, in its totality, makes up the artwork.[48]

Langlois's Softimage was only one of a group of Canadian animation application software businesses to reach widespread international adoption. Alias, in Toronto, was founded in 1985 by entrepreneurs with a background in graphics effects services, after they realized that there was money to be made building software and selling copies to smaller effects houses unable to support in-house custom development. The initial customer base for Alias used it mainly for 3D modeling, and early important clients included firms based in the Detroit automobile industry; its forte was not animation at all at the outset. The academic environments of the Universities of Toronto and Waterloo were important in providing skilled talent and resources to help Alias's advance.[49] Softimage was Canadian-held only for the first decade of its corporate life: Microsoft acquired Softimage in 1994. AutoDesk acquired Alias in 2004 and built on its code base to support the still industry-standard program, Maya.

Side Effects, another Toronto animation software company, was cofounded by Greg Hermanovic, one of the electrical engineers influenced by McLaren as seen in chapter 2. This firm remains Canadian-based and controlled, and it maintains a distinct niche for "procedural" computer tools, which broadly speaking support algorithmic rather than manual methods. Hermanovic's initial inspirations for this aspect of Side Effects's core technologies were informed by performing electronic music and making experimental animation productions. That is, he was attracted more by the improvisatory and performative, rather than the more deliberative and formalist, aspects of computerized art. Implementing a programming style that plugs together networked "operators" using a graphical user interface, the approach Hermanovic invented proved attractive not only in computer graphics interfaces, but also in live multimedia performance.

In comparison with parallel US efforts in computer animation, the trajectory recounted in the Canadian case was enabled by the relative stability of its initial basis in an interinstitutional collaboration involving two well-established government agencies. By contrast, John Whitney worked as a freelance artist-in-residence over a few years in the 1960s at IBM research labs and continued his research at CalTech and later in collaboration with artist-programmer Larry Cuba. As computing infrastructures shifted from the early installation phase to widespread inexpensive deployment of off-the-shelf hardware and software, Whitney became increasingly invested in promoting his particular neo-Pythagorean vision of visual music, pursuing "digital harmonies" on small systems garnering relatively weaker results than either his landmark analog computing or early digital productions.[50] VanDerBeek and Schwartz were engaged in computer animation primarily during the golden years for the experimental arts at Bell Labs, and their follow-up productions were not, as was evident in the Canadian story, generative of system-level advances.[51] Tom De Fanti and Dan Sandin's Graphics Habitat at the University of Illinois spawned a range of diffusible outputs from the 1960s on, including a powerful analog video processor, the GRASS graphics language; in the 1990s, their student Carolina Cruz-Neira would go on to design and develop the widely deployed CAVE, a practical system for generating immersive 3D virtual environments.

The core group of researchers mainly credited with the rise of computer-generated Hollywood feature film production worked together over a long trajectory from MIT's Lincoln Lab to Pixar.[52] Notably this group had very little involvement with experimental production, with the notable exception of Ed Emshwiller's 1979 *Sunstone*, made at the New York Institute of Technology in collaboration with scientist Alvy Ray Smith. This is a film, like *Hunger*, that succeeds aesthetically as well as effectively serving to reveal new imaginative futures. In *Sunstone*, the emergence of 3D navigable space from 2D color mapping ignited creative imaginations in both the technical community and the arts community.[53] But, for all its brilliance, it had only the status of a side project within the NYIT research program to reengineer the form and aesthetic content of Disney feature films.[54] The Canadian experience of early experimental computer animation, born initially out of artisanal, small-scale support for precisely the handmade cinema that was foundational to the alternative Canadian technological ethos over the same years, was more modest, but not less generative in its formative effect on the emergence of a homegrown digital culture.

4 Calculated Motions: Gesture and Expression at the Interface

The alternative technological ethos presented in previous chapters, employing sensorial immediacy and improvisational expressiveness to explore the properties of emerging media technologies, comes into direct focus in this chapter on the art, science, and critique of the human-machine interface.

The term "interface" entered common parlance during the 1960s, conveying a dual set of meanings signifying both object and process, apparatus and activity. Just as in the early modern period, when "philosophical instruments" played at the margins between science, magic, entertainment, and inquiry,[1] so the interface is a boundary concept crucial for understanding another period of turbulence and rapid change: the transitional decades into the information age. During this transitional period—roughly between 1970 and 1990—Canadian experiments in the arts and media emphasized the interface, and were carried out by artists, scientists, and engineers in ways that also bridged the boundaries between the conventional functions of the studio and the laboratory. The cases explored here span computer animation, electronic music, experimental cinema, kinetic sculpture, interactive sound environments, and computer-aided choreography. In all, the interface came to be understood as a negotiable zone for balancing between programmed control and expressive freedom, between the capture and playback of fixed movements and the use of improvisatory instruments for spontaneous expression. The creative arts played a catalytic role in the exploration of the interface in Canada, and they also served to spur forward the field of human-computer interaction internationally, as a new interdisciplinary terrain spanning science, engineering design, and the arts.

For both senses of the term interface—as object and as process—it is well to recall that Canada itself has been characterized as the "borderline case."

This was the title of Marshall McLuhan's 1967 essay initially commissioned by the CBC as a centennial radio broadcast. One effect of Canada's continent-wide unfortified border with the United States, McLuhan argued, is to nourish a "perpetual philosophic mood" favoring flexibility and dialogue. This Canadian "between-ness" is "the world of the interval, the borderline, the interface of worlds and situations. . . . *The interface is where the action is.* No need to move or follow, but only to tune perceptions on the spot."[2] In a similar vein, writing of computing and media, contemporary communication theorist Alex Galloway insists on the interface as an "active threshold mediating between two states," a state of "being on the boundary. . . . The interface is not a thing, an interface is always an effect. It is always a process or a translation." More than a technical location where humans and machines meet, the interface is not simple and transparent but a "fertile nexus."[3] Transposed to the circumstances of Canada as nation, this sense of the "fertile nexus" of the interface describes the Canadian context at numerous scales: geopolitical, institutional, disciplinary, interpersonal, as well as at the scale of the various kinds of objects under consideration here—instruments, screens, control devices, sensors, software, and artworks.

When McLuhan understood the borderline condition as a kind of interface—"where the action is"—he was emphasizing its potential to serve as an amplifier and creative resource, producing resonant energies from its varying mixture of proximity and distance. Jody Berland has expanded on this duality of perspective, arguing that the Canadian is capable of "'passing' as the other while recognizing the other as not himself."[4] When Canadian cultural and scientific institutions are able to function dually as "studio" and "laboratory," as seen in previous chapters in the cases of both the NFB and NRC, they can assume the appearance of "an interface between art, science and industry."[5] As Berland points out, "social and cultural studies of technology concur that we share our human agency with keyboards, software, and implants."[6] Indeed, critical voices in the field of human-computer interaction prefer to consider the interface as a mutual process of reconfiguration, not a rigid, essentially unwavering condition.[7]

At the scale of the interface as object, the commonly assumed idea, conditioned by the widespread acceptance of the graphical user interface for the personal computer, has long been that interfaces should be easy-to-learn devices that "blend with the task" in order to "ease the burden on the user."[8] Yet such a potentially dumbed-down notion of the interface has been often critiqued over the past half-century. Early researchers, like Douglas Engelbart during the 1960s, thought of the user as acquiring progressively more powerful kinesthetic and motor skills: in effect, operating interfaces with greater instrumental virtuosity to keep pace with the mental scope and expressive boundaries set by the user's intellect and kinesthetic skills.[9] Alan Kay, who headed the research at Xerox PARC leading to the first personal computers, has long railed against what he considers to be mistakenly facile interfaces that disable positive learning curves, as opposed to interfaces designed with a measured balance of challenge and reward that pay off progressively over time and allow users to become experts.[10] Exactly this sort of learning arises in the various scenes of artist-designer collaboration in the cases that follow. Each one drew together the creative and inventive talents of multidisciplinary teams or individuals. In some cases, direct manual approaches were transposed from established creative practices (film animation, music) to serve the needs of animators, musicians, or choreographers to conceptualize or compose complex sequences in real time that would be hard to do otherwise. The last case showcases an artist-engineer who pursued the interface as the basis of a new genre of art itself: interactive sound sculpture.

Beginning in the 1960s at the height of the Cold War technological arms race, when practical needs urged new methods of interfacing humans with computers, the chapter tracks the interface through musical studios and engineering labs, dance choreography workshops, hacked camera controllers on film locations, and the workspaces of Canada's earliest interactive artists. By the end of the period covered, around 1990, interfacing with computers was starting to become the everyday, indeed banal fixture of ordinary life it is now; the critical stances taken even then toward the complexity of interface as object and as process are still instructive today.

Art Animation at Lincoln Lab: Knowledge Transfers between Studio and Laboratory

The groundwork for interactive computer graphics occurred at the Lincoln Laboratory of the Massachusetts Institute of Technology during the 1950s. Driven by the SAGE project, a massive military program for air defense and automated battle control, the Lincoln Laboratory research aimed at supporting real-time surveillance of North American airspace, which became the first time a computer was used to control a large geographically distributed system. In the process, unprecedented graphic interaction techniques were developed. These included display monitors (then called scopes) and pointing devices (called light guns) permitting the selection of virtual objects. Out of this technological foundation, within a top-secret military base in the late 1950s, a new approach to human-computer interaction emerged, built around the research platform of the TX-2 computer. Fabled in the lore of computing history, the TX-2 had 70,000 words of memory, more than twice as much as other large machines of the day, and occupied a medium-sized room. The TX-2 had every known gadget for interaction—switches, knobs, push buttons, tablet, light pen, and display.[11]

By the early 1960s, a broader civilian generalization of real-time computing had grown up at MIT and spread to the wider Cambridge environs. These more general implications had been appreciated at an early date by psychologist J. R. C. Licklider, whose 1960 article "Man-Machine Symbiosis" was closely informed by his and his circle's concrete experiences with the earliest real-time graphics systems at Lincoln Lab. In this influential text, Licklider articulates with striking clarity the ideal of a symbiotic partnership between people and computers, mediated by software and interface devices to facilitate "formulative" thinking. What he termed the "third space"—an intermediate zone of symbiosis—would, Licklider forecast, come to "constitute a dynamic, moldable medium that can revolutionize the art of modeling. Trial-and-error procedure . . . [would] permit non-programmers to use the computer to help formulate the terms of new problems, not just use it to solve problems already formulated."[12] As Paul Edwards has noted, Licklider's article rapidly achieved the status of a reference point in computer science, "the universally cited founding articulation of the movement to establish a time-sharing, interactive computing regime."[13]

In terms of technical advances on the TX-2, the most influential was Ivan Sutherland's Sketchpad, "an interactive design tool for the creation, manipulation, and display of geometric objects in two or three-dimensional space."[14] Sutherland's MIT doctoral project, often called the most influential thesis in the history of computer science, provided the foundation not only for a new industry (computer-aided design), but a new conceptual paradigm for how interactive computers could be imagined. Sketchpad gave users the ability to make stick figure designs using a light pen drawing directly on the face of the monitor, to modify positions and sizes, to duplicate and paste them together, all interactively in service of an evolving design process.

Prominent among the next crop of results after Sketchpad was a system for "interactive computer mediated animation," implemented as a dissertation project in electrical engineering at MIT by Ronald Baecker between 1966 and 1969. In interviews conducted for this book, Baecker acknowledged that he was greatly influenced by the breakthroughs in user interactivity demonstrated by Sutherland's Sketchpad, and by Licklider's vision of human-computer symbiosis.[15]

While working on his dissertation during the mid-1960s, Baecker was drawn to and inspired by the experimental film culture of Harvard University's Carpenter Center for the Visual Arts. Some of Harvard's faculty were former National Film Board animators or would go on to join the NFB. Baecker was initially just curious about the use of computers to draw figures, having already experimented with them on a time-sharing mainframe at MIT. He was lucky to have as an early advisor Seymour Papert, the influential mathematician and computer theorist, who devised LOGO, an early language for teaching children to program computers. Papert and Baecker agreed that Baecker would start trying to use LOGO to animate figures, and he succeeded in making a few stick figures walk on a timesharing system in 1967.

At the Carpenter Center, Baecker was particularly impressed by the range of experimental techniques and creativity found in the work of McLaren and other NFB animators, in which he saw "incredible richness of motion control; simple motion over complexity of imagery." Baecker supported McLaren's definition of animation as

> not the art of drawings that move but the art of movements that are drawn. What happens between each frame is more important than what exists on each frame. Animation is therefore the art of manipulating the invisible interstices that lie between the frames. The interstices are the bones, flesh and blood of the movie, what is on each frame, merely the clothing.[16]

McLaren's oft-cited definition indicates the clarity with which Baecker formulated the conceptual challenge of animation in terms of pure motion dynamics—not the surface rendering of detailed textures or the anthropocentric expressive acting of Disney-style feature film animation.

Considering the diverse forms of creative filmmaking he saw on offer at Harvard, by 1967 Baecker started to think, "What could you do to couple an artist directly to the computer?" In that year he was invited to show a simple animated stick figure at the New York chapter of Experiments in Art and Technology, an ongoing forum for meetings between artists and Bell

Labs engineers following that organization's fabled series of performances *9 Evenings: Theatre and Engineering* in 1966. At the Experiments in Art and Technology meeting, Baecker met Michael Noll, a Bell Labs researcher who had done some early animations for scientific visualization. Noll introduced Baecker to Eric Martin, an animator then teaching at Harvard. Martin would later express his own artistic credo in these terms: "Animation is the graphic art which occurs in time. Whereas a static image (such as a Picasso etching or complex graph) may convey complex information through a single picture, animation conveys equally complex information through a sequence of images seen in time."[17]

Broadening his reflection on film aesthetics, Martin pointed out that the changes between any two forms, considered in terms of information quantity, were slight. But the key factor in conveying mood or emotion is implicit in picture *change*—shifts of position, shape, or dynamics.

> Therefore, a computer is ideally suited to making animation "possible" through the fluid refinement of these changes. . . . Imagine a man walking in the dead of night. The only evidence of his presence is a set of luminous points attached at random to his body. At any instant, we are aware of these points, but not their significance. In motion, we read a walking man as the synthesis.[18]

Martin worked with Baecker at Lincoln Lab every Sunday for more than a year in 1967–1968. His role was to use Baecker's roughly mocked-up animation system for drawing motion trajectories, to provide suggestions for new functions, and to evaluate these functions as they came online. After his second or third visit to Lincoln Lab, Martin came to the realization that this was "a radically different medium." Martin recounts,

> The imagery was very limited. We'd draw a ball with seven dots. We both strongly agreed that what was important was not what was in the frame, but what was in the behavior. Ron was happy to put aside pictorial detail, and to focus on motion. This is what I realized was radically unique, and it has to just come to you, this kind of realization—the idea of motion as distinct entity. Before, in traditional animation technology, motion wasn't a separate thing, it was implicit, embedded in the images as such. The light bulb went on, how it went beyond the constraints of previous film animation. I asked Ron, would it be possible to create a motion that would have its own integrity, that would be a separate thing? The next week, Ron had implemented it. It was possible to draw a path.[19]

Baecker expressed the close collaboration in similar terms:

> I was very much influenced by Eric, and the minimalist school of animation. Looking at what you can do with three dots—if three dots move well, on that you

can base a story. . . . My insight was that if you could draw images you could also draw motions. Putting all that together became the Genesys package and picture-driven animation, operational by summer 1968. (Genesys is an abbreviation for GENEralized cel animation SYStem.)

Designated as P-curves (parameter curves) in Baecker's thesis, this contribution permitted, in a later textbook account, the "substitution of a visual program for a textual one: rather than explicitly writing out descriptions of actions, the animator provides a picture of the action. A P-curve is a parametric representation of the motion (or any other attribute) of an object or assembly of objects in a scene. The animator describes an object path by graphically specifying its coordinates as a function of time."[20] Baecker also stressed the novel power of trial-and-error experimentation, not available on film:

The power of immediate visual feedback is striking. . . . The animator can see and not merely imagine the result of varying the movement and the rhythm of a dynamic display. He will be led to perfect that aspect of animation that is its core: control of the changing spatial and temporal relationships of graphic information.[21]

Baecker was able to hire an artist to be filmed demonstrating the system. The artist, Lynn Smith, recounts that after a brief introduction she was able to work on her own, using Baecker and Martin's Genesys system to produce an unauthorized film working at nights when the lab was quiet and there would be little attention to the fact that the equipment was being used by a civilian without clearance. As an artist improbably working at Lincoln Lab, one of the highest-security military bases in the country, she had little idea what was going on there; nor did higher authorities ever catch on to what she was doing.[22] Smith's film was shown by Baecker at meetings and conferences, spreading the reputation of the Genesys system within the growing computing research community of the early 1970s, a community that was starting to become diversified in both social and cultural terms. For Ted Nelson, the enthusiastic countercultural promoter of graphics computers as "Dream Machines," the basic elegance of Baecker's system was its simplicity: "it made everything work the same way, through control by screen diagrams. [Baecker] had simplified the animation problem in a clear and simple way."[23]

Baecker immigrated to Canada for his first academic job at the University of Toronto's Department of Computer Science, after receiving an invitation from Leslie Mezei, an expert in and practitioner of computer art. This department had been founded in the 1950s by Carl (Kelly) Gotlieb, and was the first established in Canada. Gotlieb's interests were wide-ranging, including considerations of the human and social impacts of computing,[24] and he was responsible for hiring Mezei as a faculty member, despite Mezei's lack of advanced degrees in computer science. The department under Gotlieb's leadership would go on to welcome much broader exploration of the intersections between the arts and computing, and also social informatics and user-centered design.[25]

A Hungarian Holocaust survivor, Mezei had worked as an actuary before entering academia in 1966, initially as a participant in the interdisciplinary seminar led by McLuhan student Donald Theall at Toronto's recently established York University. Mezei recalled that in those years, "we imagined a new Renaissance, a great flowering of the arts," inspired in part by the 1967 world's fair in Montréal, a climate that was nurtured too by the ongoing presence of McLuhan and others at the older University of Toronto.[26] As a professor with interdisciplinary interests in a department led by a chair encouraging the exploration of computing in relation to human values, Mezei was not only one of the first to write serious criticism about computing in the visual arts,[27] but also established a research project himself in 1966 for creating algorithmically generated images using an IBM mainframe. This fledgling computer art research program developed in parallel, though at a much smaller scale, with projects at US research universities that aimed to train aesthetically minded "engineers for change." Most notably, a center for "advanced visual studies" was founded at MIT that institutionalized what one scholar has called "aesthetic virtue in the Postwar Defense Institute."[28] Mezei, like Kar Liang in the same years at the National Film Board, closely studied what was going on across the United States and was most impressed by Ron Baecker's Genesys program, admiring in particular its gestural control of motion paths. Baecker initially joined the Toronto department as a sabbatical replacement for Mezei, then as a regular appointment.

In 1973 Baecker's work also attracted the interest of Alan Kay at Xerox Palo Alto Research Center (PARC). Founded in 1970, PARC is now acknowledged as the site of several of the key innovations of the modern computer, including Ethernet, laser printing, and, centrally for our present interests, the now-standard graphical user interface for the personal computer.[29] Alan Kay was a versatile and charismatic scientist who had been a professional musician and earned a PhD in computer science from the University of Utah, where Ivan Sutherland among others had relocated the leading edge of graphics research. By 1974, Kay's Learning Systems Group at PARC was well along in assembling the basic hardware and software components for what would become the personal computer. This work built on several sources, notably Sutherland's Sketchpad, Douglas Engelbart's work on interfaces at the Stanford Research Institute, and Seymour Papert's research at MIT on end-user programming. Kay was also an avid reader of Marshall McLuhan's works, which informed his thinking on new media as "environments," quoting with enthusiasm McLuhan's adage that "once we have shaped our tools, the tools turn around and reshape us."[30] Kay was interested in supporting a research project on animation as one of several possible applications of the new Alto personal computer prototype and its programming language, SmallTalk, which was at that point still untested by real users. The interface on the Alto included what would become the standard graphic interface: icons, windows, menus, and a mouse for selection and manipulation of objects.[31]

Under Kay's sponsorship, Baecker put together a small project to reimplement and improve Genesys in SmallTalk. His three-person group included Eric Martin and a graduate student programmer from Toronto. Developed over two summers at PARC, in 1974 and 1975, Baecker's new program, rechristened Shazam, was a key example of what Kay would later discuss as "Personal Dynamic Media" in a now-classic article by this title. Kay's driving vision was of a creative computer literacy that combined the discipline of explicit text-based symbolic programming with an intuitive iconic direct manipulation of visual objects. For Kay and his associates, Shazam served as an ideal illustration of what interactive computing could become:

> Animation, music and programming can be thought of as different *sensory views* of dynamic processes. The structural similarities among them are apparent in SmallTalk, which provides a common framework for expressing these ideas.[32]

Two details of the development of Genesys in the PARC environment are worth underlining here. The first is the notion of displaying two simultaneously active windows with several different views of the same dynamic process. While one window was showing a looping preview of the animated object, the user could operate on it in real time in the other window.[33] Second was the clear evidence of Kay's ambition that SmallTalk might serve to program applications designed to have a "low floor," allowing quick entry access by artists, as well as a "high ceiling," that is, providing room for the addition of new features and system extensions as users' skills develop. Thus, in a 1976 paper, Baecker would write that "a suitably skilled animator may himself determine aspects of his animation *system*, only if the system is not a fixed set of commands but an extensible, truly open-ended programming language."[34]

While Shazam was a relatively small program, the most developed of the early SmallTalk projects, according to Kay, was David Smith's Stanford dissertation project, Pygmalion, "an essay into iconic programming."[35] For Smith, Baecker's interactive animation system was a prime demonstration that programming a computer need not be limited to textual coding: Baecker's innovation was, Smith wrote, at "the core of Pygmalion's model of articulate communication."[36]

Back in Toronto, Baecker and his students extended their work into computer graphics more broadly, all the while maintaining a close interest in the problems of interaction. By the mid-1970s, meanwhile, Mezei grew disillusioned with the promise of computing in general, partly due to the growing technical narrowness that came with increasing professionalization of the discipline within the department, partly due to what he considered to be a focus on technology for its own sake. According to his German colleague Frieder Nake, Mezei felt that "issues of humanity are more important than those of technology. Art without a clear relation to the conditions of human existence was not his concern. His hopes of ten years earlier did not ripen. So, he quit."[37] But not before bringing Bill Buxton, a young musician with experience in electronic and computer music, into the department.[38] Together Baecker and Buxton, both informed as researchers by the challenging demands of designing creative interfaces for use by artists, would go on to becoming founding figures in the emergent interdiscipline of human-computer interaction (HCI), spawning generations of Canadian researchers in the field.

Musical Instruments and the Playable Interface

While still an undergraduate music student at Queen's University, Bill Buxton was sent by his composition teacher to visit the National Research Centre in Ottawa, where he was able to explore the musical component of the same research system that was used to support the production of *Hunger* by Peter Foldes at the NFB (see chapter 3). At the NRC Buxton was shown and given instruction in using the early interactive music system built there by Ken Pulfer. Pulfer was a computer systems engineer and amateur musician who began his career at the NRC during World War II working on applications of radar. During the early 1960s, he switched his interests to upper atmosphere rocket instrumentation, and then to computing. With a background in rocket hardware control, Pulfer set the direction for his interactive music research when he selected an industrial control minicomputer as the platform for a new research program in "man-machine communication for creative applications";[39] a real-time control computer was chosen for its speed and responsiveness. He was "not interested in calculation per se, but in problem solving in the general sense."[40] Therefore, the relation between the computer and the outside world was as important to him as what went on inside it. From the start, the question that interested Pulfer was the use of the computer by "non-literate" people, and he chose music composition and performance as the application to develop new general understanding.[41] He widened the scope of his team working on interactive applications when he brought Marceli Wein and Nestor Burtnyk to join him, working separately on computer animation (see chapter 3).

As Pulfer surveyed the state of computer music in the 1960s, he saw little attention being paid to the problem of interaction, so chose this as his area of focus. In part this arose out of his own practically oriented interest in supporting the needs of working musicians. Pulfer's approach led him eventually to open the lab doors wide to musicians working across stylistic divides, from commercial film composers to academic avant-garde experimentalists. The system he designed supported music composition and editing, with timbral control and on-screen notation display, a piano-like keyboard for data entry, plus a chord keyset and mouse akin to the devices just invented by Douglas Engelbart in California.

Pulfer may also have been influenced by the approach being taken by his colleague at the NRC in Ottawa, Dr. Hugh Le Caine, a nuclear physicist turned instrument inventor who is credited with creating the first electronic synthesizers, in research dating back to 1945. Over the course of three decades, Le Caine's design emphasis was on the "playability" of his electronic instruments, from his first invention, the Electronic Sackbut, to a polyphonic variable-speed tape recorder and various kinds of touch surfaces.[42] The Sackbut, for instance, offered intuitive touch control over the sound quality as well as pitch and amplitude. The variable-speed tape recorder allowed for keyboard control of up to ten parallel channels of transposing tape loops. Le Caine worked closely with Canadian university music departments, particularly at Queen's, McGill, and Toronto, supplying them with instruments and responding to their requests. Both experimental and commercial composers considered Le Caine to be a sophisticated musician with deep insight into the psychological aspects of creative exploration in music composition. Israeli composer Joseph Tal was one of the first to take delivery of the special-purpose tape recorder, which he credited with "reducing the span of time between thought and action. This was an instrument with a 'mind' aligned to that of the composer, an ingenious construction because of the rare combination of technique and imagination."[43]

To describe a musical instrument as having a "mind," as Tal did, is not merely a figure of speech. Both anthropologists and experimental musicians have written of instruments as possessing a degree of independent agency.[44] "Playability," the value emphasized by Le Caine, is not just a matter of simple efficacy, but a two-way dialogue in which the materiality of instrumentation excites new modes of musical thinking, and new thinking helps inform new kinds of instruments.[45] Buxton would go on to transpose Le Caine's method of designing for and with users into a foundational precept for HCI, particularly with respect to touch sensitivity. Indeed, with tongue only partly in cheek, Buxton would come to call the specification of interfaces for the exigent uses of artists the most challenging of tasks for designers: "No profession has a longer history of articulating powerful and creative ideas through a technological intermediary than musicians."[46] In a similar vein, another musical computer scientist, Jaron Lanier, has lately written that not only are musical instruments "the best haptic interfaces yet invented, they are the best interfaces of any kind, if what we care about is the potential for mastery and expressiveness. Instruments show what's possible, how far computer science must go before we call it even a beginning."[47]

The system for music composition devised by Ken Pulfer was, for Buxton, a "beautiful music machine" allowing for bimanual input and a host of other features that made it the functional equivalent of the personal computers that would only become widely available two decades later. Its research purpose was not musical per se, but the study of user-computer interaction through the solution of musical problems—musicality as an instance of the general case of creative interaction.[48]

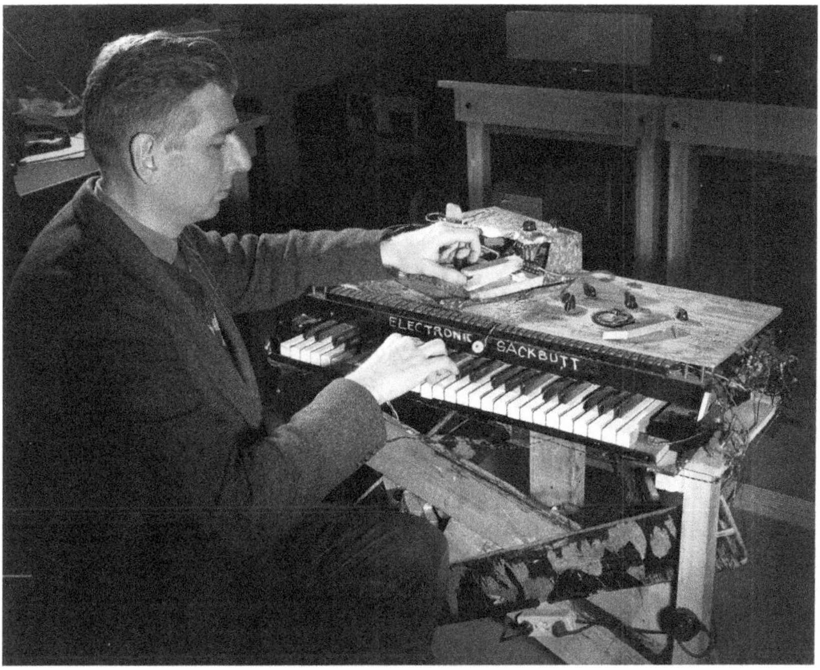

Figure 4.1
Hugh Le Caine with the Electronic Sackbut, ca. 1950s. Courtesy of the National Research Council.

Following these two formative experiences in Ottawa, Buxton went to a more formal computer music lab at the Institute of Sonology in Utrecht, Holland, where he followed in the footsteps of another Canadian composer, Barry Truax (who will be a key figure in the next chapter). There, Buxton stumbled upon algorithmically produced images by Leslie Mezei in the catalog of one of the signature art and technology exhibitions of the era, *Cybernetic Serendipity*.[49] After making contact with Mezei and receiving his encouragement, Buxton joined Toronto's Computer Science Department in the mid-1970s, initially as a graduate student and then as the initiator of his own externally funded computer music research program emphasizing real-time performative capabilities—the Structured Sound Synthesis Project (SSSP). The SSSP (1977–1985) was a full-blown interactive computer-assisted music composition system, including means for polyphonic digital synthesis, score editing, and real-time performance controllers.[50]

In one of the early publications coming from the SSSP, Buxton theorized music performance systems as a three-dimensional space.[51] Figure 4.2, derived from that study, summarizes these three dimensions. The horizontal dimension presents a continuum between the replay of prestored fixed sequences and the spontaneous, improvisatory generation of new sequences. This is called the amount of real-time decision making afforded by the system. Along the vertical dimension lies the degree of complexity that can be controlled by the system, from individual instances like notes to larger-scale sections or patterns. This is called the "structural level" supported by the system, an indicator of the scale of musical material that can be affected by the system. The third dimension is defined as the distribution of agency "between musician and machine." This continuum is between a single locus of control (which could be *either* human or machine) and a divided manner in which decision making between human and machine is shared. The theoretical insight derived from music composition and performance can

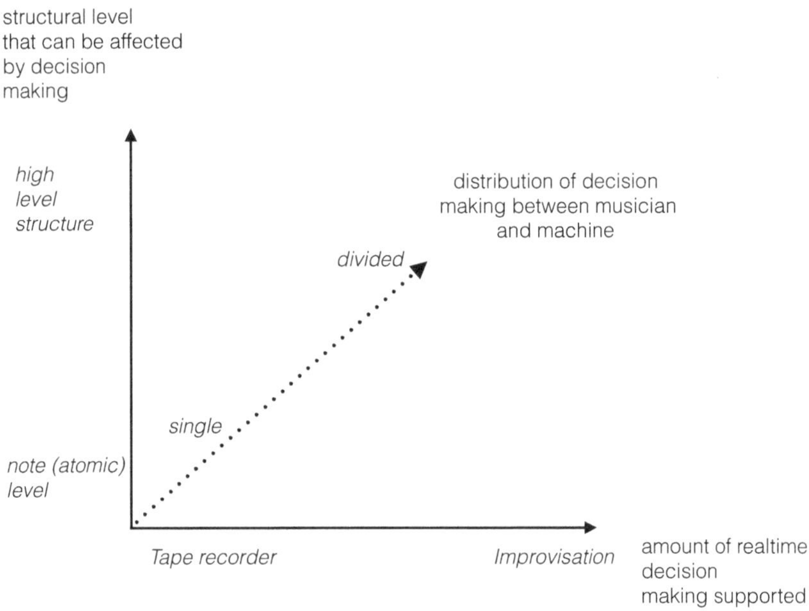

Figure 4.2
Three-dimensional space used to characterize performance systems, adapted from Buxton et al., "A Microcomputer-Based Conducting System," *Computer Music Journal* 4, no. 1 (1980).

be broadened into a general theory of interactive systems, and indicates at an early stage what would grow into Buxton's approach as a leading HCI theorist as well as researcher and designer. In the specific instance of the Conduct program within the SSSP toolset, Buxton insisted on building an instrument to enable the musician to improvise and adapt performances to different venues and audiences, at a time when computer music was almost always presented in a fixed recorded format for playback only, whether on tape alone or with live musicians playing along.

Funding for this research came from Canada's Natural Sciences and Engineering Research Council, not from arts sources. As a result of the interests of these funders, Buxton increasingly found himself needing to develop a doubly coded language for talking about and justifying the role of musical "problem solving" in a research context, starting with the generic name of the project, Structured Sound Synthesis, which avoids any mention of music. Buxton avoided aesthetically tinged terms, like musical creation, improvisation, and expression, for the purpose of reporting on this work as engineering research, instead using a vocabulary based in terms like problem solving, user-modeling, constraints, and affordances.

Responding to the request of a professional percussionist in the renowned Nexus ensemble, Buxton and co-researchers also implemented one of the earliest examples of multi-touch interaction, now commonplace on mobile and tablet devices. This invention was initially for the purpose of controlling an electronic drum. Multi-touch sensing, so typical in musical instrument performance, was described as a method of continuous adjustment of control parameters.[52] Buxton, like Baecker earlier, was soon invited to be a visiting researcher at Xerox PARC to work on interface concepts and designs for what would become ubiquitous computing devices, and after a stint as a research director at Alias, a Toronto graphics and animation software company, became a senior scientist at Microsoft, where he contributed to the conception of commercial tablets employing multi-touch interaction. In recent years, Buxton attributed signal importance to his early years as a musician for his later eminence as a founding figure in the field of HCI. Referring to his foundational work on touch interfaces, he recalled, "I wasn't trying to make a computer interface, I was just trying to make a drum. Did I envision what was going to happen today, that it would be in everybody's pocket—in their smartphone? Absolutely not. Did we realize that things were going to be different, that you could do things that we never imagined? Absolutely."[53]

Along with Baecker, Buxton was a key figure in contributing to a wide-spanning intellectual foundation for HCI, with the pair jointly editing two comprehensive, self-consciously discipline-forming scientific volumes on the topic in 1989 and 1995.[54] Both volumes begin with a coauthored essay providing an early version of what has become the canonic genealogy of interaction philosophies and milestones. Between the two collections, separated by only six years, a vast expansion of the field occurred, necessitating a nearly complete revision of the first volume's framework to include cyberspace (the early internet), virtual reality, and mobile and ubiquitous computing.

LifeForms: Choreographing the Polyrhythmic Body

At Simon Fraser University (SFU) in Vancouver, founded in 1965 and imbued with the spirit of an upstart campus born at the height of 1960s radicalism, a Faculty of Interdisciplinary Studies provided the home for an unusual conjuncture of computer science, communication studies, contemporary arts, and—what might initially appear the odd element out—kinesiology. While seemingly disparate, each of these departments was open, venturesome, and entrepreneurial, in the recollection of one of the Faculty's early and formative deans, Tom Calvert.[55] Calvert was a computer scientist, with an early interest in computerizing movement through notation schemes for applications across various domains beyond dance itself—industrial time and motion studies, kinesiology, and animation. Building on a notation scheme devised in 1928 by Rudolf Laban, called *Schrifttanz* ("written dance"), Calvert's research developed a conceptual model that considered the moving body as "analogous to a multiprocessor," allowing each body segment to be specified independently.[56] By the early 1980s, working with a graduate student in kinesiology, Calvert laid the foundations for what became LifeForms, one of the first software applications for composing movements with an interface designed for use by dancers and choreographers.

Writing dance, to use Laban's initial phrase, has a much shorter history than the highly elaborated, thousand-year lineage of Western music notation, a practice that, as Friedrich Kittler noted, is one of the oldest means of "capturing time."[57] As the musicologist and composer Hugues Dufourt has suggested, notation in this way "arrests time . . . interrupts its traditional free flow and attachment to conventional embodied practices."[58] Music notation has served too as the basis for a comprehensive theory of symbolization in art, published in Nelson Goodman's 1968 treatise *Languages of Art*. For Goodman, standard music notation indeed captures time in a unique way, permitting the definition of musical works independently of the variety of actual interpretations with which they might be played. Goodman's theory applies primarily to conventional repertoire, since, by the mid-twentieth century, experimental composers had begun to devise nonstandard schemes to notate their works. Often inventing graphical devices proper to each individual work, experimental notations typically offer a suggestive pattern rather than a deterministic script for a performance.[59]

With LifeForms, Calvert and his subsequent collaborators also made possible a kind of gestural "writing of the body" that goes beyond the function of notational capture, analysis, and playback, to afford creative artists the capacity to expand the expressive limits of what the body can do.[60] Thecla Schiphorst, a dancer with background in computer programming at SFU, worked with choreographer Merce Cunningham on a series of choreographic compositions that aimed to extend the capabilities of the LifeForms software. These works were later toured extensively by Cunningham's company.

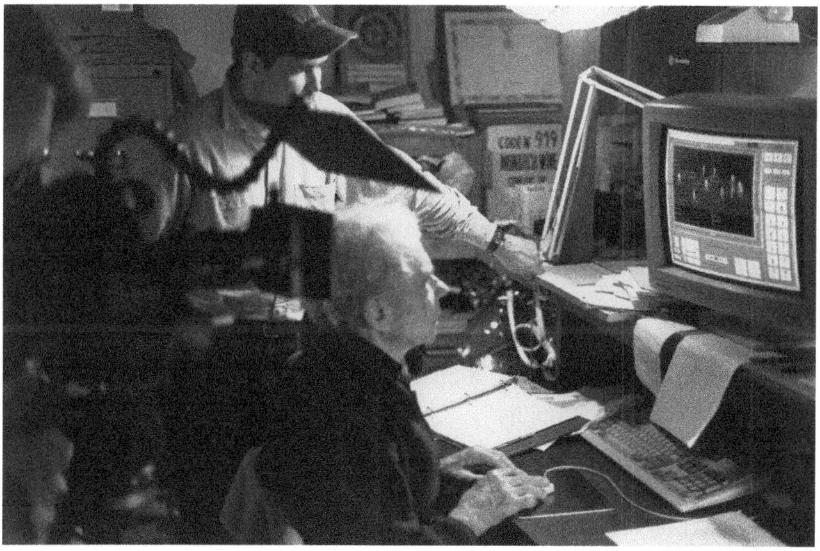

Figure 4.3
Merce Cunningham (foreground) working with the computer software LifeForms during the making of Elliot Caplan's (background) documentary *CRWDSPCR*. Photographed by Michael Stier, 1993. Courtesy of the Merce Cunningham Trust.

Cunningham had a long-standing interest in using computers as a creative instrument, having imagined as early as 1968 an electronic three-dimensional notation for visualizing and editing shapes and timing of skeletal figures.[61] Cunningham also had originated an extensive practice using chance procedures, applying the same methods to dance that composer John Cage had introduced to music composition in the early 1950s. For Cunningham, chance procedures put him in touch with "a natural resource far greater than my own personal inventiveness could ever be, much more universally human than the particular habits of my own practice."[62] Chance, in other words, served to free Cunningham's choreographic imagination from the formulaic entrained motions of corporeal routine. Both Cage and Cunningham used chance procedures in ways that, according to media artist and computer scientist Marc Downie, had

> a flavor of computation and transformative representation about them. Motions were broken up, atoms identified, tokenized. These arbitrary tokens rearranged by the toss of a coin, the fall of the I-Ching, and the new lines and tables recast the new motion material for his dancers. . . . Their quest for new compositional strategies could be re-read as a quest for representations that support useful compositional actions.[63]

Throughout an iterative design process, the team working with Cunningham considered him as a "mentor user," whose wide experience and creativity could drive such system-level innovation. Under Schiphorst's guidance, Cunningham began working with LifeForms using just its initial palette, which offered the ordering, editing, and animating of individual body stances. In response to his input, the development team at SFU modified the system to make a sequence of dance phrases the basic building block for the choreographer. Other modifications included new graphic controls, including "inverse kinematics," which allowed the user to select and modify the endpoint of a chain of body parts rather than having to move each part individually.

Treating LifeForms, therefore, as a "visual idea generator," rather than just as interface to dance notation, Cunningham's use of the developing system also served to validate what Calvert referred to as "general knowledge" of the creative design process. For instance, the provision of support for thinking at multiple levels allows for hierarchical organization, from high-level conceptual thinking to control over low-level details. Moreover, the system was reflexively understood to support creativity by offering multiple views of the stage and dancers and supplementing the "internal knowledge" of the user, with both explicit knowledge (prestored stances, libraries, sequences) and procedural knowledge (generative sequences based on algorithms). Most importantly, Calvert concluded, the iterative development of LifeForms supported advances both in the system itself and in the choreographer's creative methods.[64]

The most salient of Cunningham's innovations was, according to Schiphorst, in the area of meter and rhythm, which the choreographer approached by setting different rhythms to different parts of the body. Throughout his work with Cage and other experimental composers, Cunningham had taken an unmeasured approach to the alignment of movement phrases to musical beats. With *Trackers* (1991), the first of the dances using ideas actually generated specifically through LifeForms, Cunningham created passages using a "complex polyrhythm with legs moving irregularly, and then arms moving in their own irregular rhythmic pattern and shape."[65] In this way, the dance interface enabled expansion of corporeal capacities, which in turn could become the basis for unprecedented complex vocabularies for new choreography. In something of a reversal of the vector from improvised action to fixed inscription seen frequently in this book, the digital composition interface served in Cunningham's experimental usage to make improbable, even counterintuitive forms of motion come to life.

Michael Snow's *La Région Centrale* as Instrumental Cinema

Michael Snow (b. 1928), the Toronto-born polyartist whose output spans painting, photography, cinema, video, and free improvisation on piano and trumpet, moved to New York City in the early 1960s, where he gained a worldwide renown in the experimental genre known as structural film. This is a "cinema of structure in which the shape of the whole film is predetermined and simplified, and it is that shape which is the primal impression of the film."[66] The author of this definition, critic and historian P. Adams Sitney, called Snow the "dean" of structural film, a distinction initially based on *Wavelength* (1967), a 45-minute work comprised entirely of a continuous zoom set within an otherwise nearly static interior scene. Critical reception of *Wavelength* was effusive and tended toward the philosophical, interpreting its concern for the intensive unfolding of time as a "metaphor for consciousness itself."[67] Snow followed *Wavelength* with a work identified simply as ↔ (*Back and Forth*), defined again strictly by the camera motions interrogating a fixed location, this time made up of repeating, accelerating horizontal and vertical panning shots. As Snow noted after making ↔, "visual rhythms (in film) can have as infinitely varied qualities as rhythms in music. The camera pans continuously at various tempi, starting at a medium speed, then gradually going slower and slower, then gradually rising in speed to very, very fast."[68] Identifying camera moves as elements in permutable sets, Snow developed a structural film practice that bears comparison with other twentieth-century formal methods, such as the generalization of twelve-tone music composition known as "total serialism."

Snow realized more broadly what he was getting into in pursuing a grammar of cinema that eschewed the use of the camera as extension of the human gaze. This realization was formulated explicitly in a 1969 grant proposal to the Canadian Film Development Corporation in which Snow sought funding for his next film: the "movement of the camera as a *separate expressive entity in film is completely unexplored*." The camera, as he put it, can be an instrument, "which has expressive possibilities in itself," and is capable of having "an equal role in the film to what is being photographed," rather than being limited just to the "dictates of the story and characters being presented."[69]

Snow proposed to make a three-hour film that would generalize *all* the possibilities of camera movement, set in an austere northern landscape that would be the "equal in terms of film to the great landscape paintings of Cézanne, Poussin, Corot, Manet, Matisse, and, in Canada, the Group of Seven."[70] Snow's landscape, conceived in the year of the first lunar landing, would be an "absolute record of a piece of wilderness" that would "feel like the record of the last wilderness on earth, a film to be taken to outer space as a souvenir of what nature once was." This would be a kind of "Goodbye to Earth which I believe we are living through."[71] The funding proposal was successful, underwriting the production of *La Région Centrale*, which was shot on a remote Québec hilltop and released in 1971 to widespread and extraordinary acclaim. "Critics took it again as an axiomatic work, this time one in which Snow recapitulated all possible camera movements, those used historically and those imaginable."[72] In Canada, this reception tended to be seen as an encounter between "human sensibility" and the "terrible natural silence of the Canadian landscape," a view which accorded the film a near "mythic status." American critic John Locke understood *La Région Centrale* as epochal, for having done nothing less than changing "both your perception and appreciation of the world and of other film."[73] French critic Alain Fleischer had a more cosmic reading, seeing it as coming out of an obsession with space exploration and remote-controlled robots, providing a view of a remote inhuman planet that is, in fact, "right here within reach on earth."[74]

To enable the production, Snow designed a special camera activating machine (CAM) allowing for all possible camera motions, then commissioned Pierre Abbeloos, an engineer employed by the National Film Board in Montréal, to build it. "The camera itself is a machine, so attaching it to another personally designed machine seemed a way of augmenting its possibilities. In this case I was composing for a very special instrument. The piano is a machine too."[75] Snow scored the film according to several parallel voices: the kind of camera movement (horizontal, vertical, oblique), the degree of zoom, and speed. The score systematically goes through all permutations, speeding up to the film's frenetic, dizzying climax. As summarized by Kenneth White,

> As the film proceeds into its third hour, the apparatus turns in rapid somersaults and figure-eight patterns. Its movement crescendos, and the images blur. The scene is atomized into a flicker of brown and blue, brown and blue. At speed, the images mutate from indifferent record to pulsating abstraction.[76]

Snow credits Abbeloos for coming up with the idea of using electronic musical tones, prerecorded on audio tape, as the means for programming the CAM trajectories. These same tones were meant to provide the film's soundtrack as well, which Snow intended as a kind of further formal integration: "it would have meant the visual part was composed in terms of sound."[77]

A tradition of building mechanical devices producing sounds that correlate to visual images goes back centuries, from Abbé Castel's "ocular harpsichord" in the seventeenth century to a host of nineteenth-century optical and sonic devices.[78] Synesthetic composers and artists such as Scriabin, Kandinsky, and Schoenberg developed elaborate theories and experiments for a precinematic visual music.[79] The 1960s saw a flowering of abstract animation making "what you hear" correspond closely to "what you see," typified by McLaren's *Synchromy* (released in the same year as *La Région Centrale*). American computer graphics pioneer John Whitney published an entire treatise on computationally generated "digital harmony" positing structural

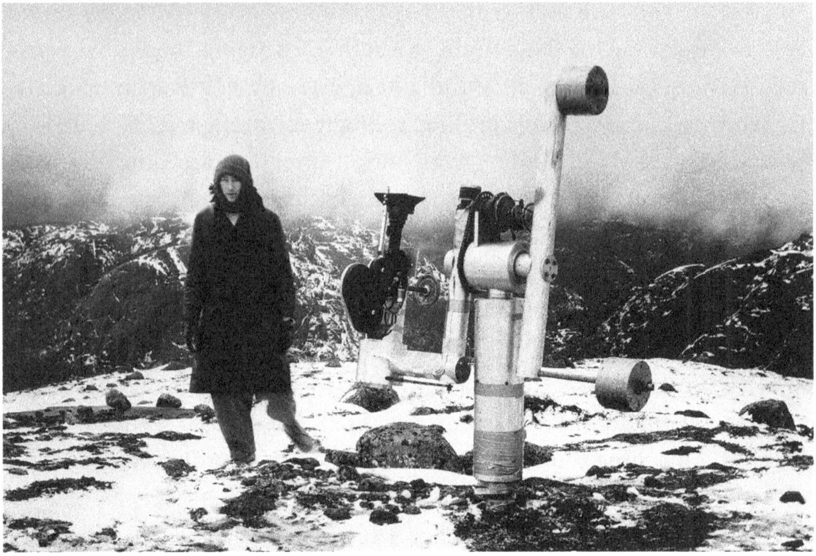

Figure 4.4
Michael Snow beside the camera activation machine in the production of *La Région Centrale*. Photograph by Joyce Wieland.

correspondences between image and sound through Pythagorean integer ratios.[80] Snow's interest in formal integration of the camera movements with the sound score and program was therefore very much of the times.

In its design, the CAM was meant to have run automatically, like a mechanical piano played using electronic tones rather than punched holes. In practice, however, the control tapes did not function correctly, and Snow ended up using a set of dials, also built by Abbeloos, to control the CAM on location. Then, in postproduction, Snow added musical tones similar to those that would have served as control signals for the soundtrack.

According to Snow, "it was a good thing that it didn't work, I had to learn to play the machine. I wrote out many versions on paper, tried to visualize. The dials had been produced beforehand by Abbeloos. I played with it, got to feel like I understood the instrument. I gained confidence, to make the moves fit the length of time available on the reel." Snow notes that toward the latter part of the film, "I had not 'used up' the faster speeds; I didn't give away the climax. It was quite an effort of concentration. . . . The whole thing suicides at the end: extreme speed kills the machine. It signifies the end of the capacity of the machine that's making it: out of control."[81] Thus, *La Région Centrale* is both structural *and* spontaneous, owing to the way Snow needed to approach the camera apparatus, in a way, according to curator Dana Friis-Hansen, that is "at once schizophrenic and cosmic, systematized and erratic."[82]

In chapter 2, I put forward an account of Norman McLaren's "instrumental cinema" through the eyes of French film critic André Martin, referring to the filmmaker's plethora of artisanal, direct-to-film and other techniques emphasizing immediacy and improvisation. This account resonates in striking ways with Snow's project in *La Région Centrale*. Snow's use of gestural expression while playing a robotically activated camera instrument allows for a reading of the *La Région Centrale* less as an austere, posthuman expression of machinic vision,[83] and more in terms consistent with what historian of science Andrew Pickering discusses as an "alternative" school of cybernetics. This alternative, mainly British rather than American approach to the science of "communication and control between man and machine" takes it as given that "the brain is an immediately embodied organ, intrinsically tied into bodily performances." Such maverick scientific figures as Gordon Pask and Stafford Beer, as well as cybernetic artist Roy Ascott, relied on this kind of open-ended experimentation. For these artists and designers, according to Pickering, "the cybernetic brain was not representational, but performative, and its role in performance was adaptation."[84] Snow's performative approach to an ostensibly structural film concept was carried out by adapting his improvisations to the affordances of the control box and the camera activating machine. In his own consideration of the control box as a quasi-instrumental device, Snow too leveraged his own refined musicianship as an accomplished improvisor.[85]

After completing *La Région Centrale*, Snow next looked at the cinematic apparatus itself as having intrinsically beautiful properties and reimagined it as a kinetic sculpture. Snow asked Abbeloos to program the CAM in a strictly determined loop with the attached camera feeding into four video monitors placed around the installation. This work, *De La* (a prepositional abbreviation for, and indicating its derivation from, the film), was made for the collection of the National Gallery in Ottawa, where it remains. Completed in 1971, it is one of the first closed-circuit video sculptures, helping at once to establish and define the genre. Here, as Snow notes, the apparatus is "fully programmed. All the possibilities are worked into a single set of moves."[86]

Sound too is important to its operation, an "essential part of the concreteness of the machine," owing to the whirring and humming of the mechanism: "if it were silent it would tend more towards a representation and also have less 'personality' as a unique Thing-in-the-world."[87] The initial intention for the use of Snow's robotic camera in *La Région Centrale*, to play back a pre-programmed set of camera trajectories, was ultimately realized in *De La* as a deterministic kinetic sculpture: motions wholly encapsulated and literally black-boxed as a discrete temporal object, analogous to a choreographic or melodic sequence of phrases.

Interactivity and Its Discontents: David Rokeby's *Very Nervous System*

Moving beyond established art forms like filmmaking or music, the next generation of artists began to consider the inherent properties of real-time interaction from an aesthetic standpoint. By the late 1970s, artists were beginning to acquire and develop skills in programming the newly available microcomputers, first using kits of components, then as configurable off-the-shelf systems. One such artist was David Rokeby, who was first introduced to component-based hardware design and programming in the experimental arts program at the Ontario College of Art (OCA) in Toronto.

OCA (now the OCAD University) in the late 1960s was shaken to its foundations by the appointment as president of a radical young British artist and educator, Roy Ascott, who brought to Toronto an experimental philosophy grounded in aesthetic concepts borrowed from cybernetics and information theory. Ascott's tenure at the OCA was tumultuous and brief, but it was formative in opening up the then-staid institution to a broad swath of approaches coming variously from conceptual and systems art, as well as bringing to the fore an openness to interdisciplinary borrowings from other media outside the visual arts. The rhetoric underpinning Ascott's manifesto for "Behaviorist Art" emphasizes the creative participation of the spectator, feedback loops, self-organizing subsystems, and properties of biological growth.[88] Ascott quickly outraged the traditional fine arts faculty and students by abolishing departments, sidelining craft-based skills like life drawing, and urging a ruthless revolutionary spirit of rebellion and iconoclasm. As president, he was seen as brilliant, impatient, and disorganized, and with his abrupt dismissal, much of the disruptive reorganization he effected reverted back to traditional structures.[89] But Ascott's flagship "Experimental Department" remained in place, which included a mix of conceptual artists, post-Cagean composers, and, by the later 1970s, artists exploring what the College designated as the "photo-electric arts." Ascott had brought in a McLuhanite creative director from the advertising business, Richard Hill, who oriented the department around the use of light and electricity as "major forms of cultural expression in the contemporary world." Hill was a polemicist, promoting, as McLuhan had broadly done, the probing role of the creative artist to make society aware of the cultural implications of these new media. In a manifesto for his department, Hill explicitly identified with "the early Renaissance concept of the artist as innovator and manifestor of ideas when the liberal arts were married to the technologies of the time." It proclaimed a technocultural future in which "microelectronics [would] return technology to human scale and . . . the individual [would be] once again important as creator, artist and human being."[90] Hill hired an American-born electronic media artist, Norman White, who had earlier participated in the activities of Experiments in Art and Technology in New York, to introduce studio courses in electronics. White considers himself to be "essentially, incurably, and unashamedly . . . a tinker. Originality, if it occurs in my work at all, is a casual by-product of my particular kind of tinkering. Tinkering is a form of reading and writing, concerned with the literature of humankind's machines."[91] It was in this climate at the OCA, based in the Experimental Department and mentored by White among others in photo-electric arts, that Rokeby established his foundational skills as hardware and software artist.

Rokeby's signature early work, and to this day one of the most widely commented-upon works in the short history of interactive installation art, is the *Very Nervous System* (1982–1991), which captures gestural expression through surveillance cameras, initially custom-built, and converts it to electronic sounds using software developed by the artist. Sound, for Rokeby, exists in the installation as a kind of "kinetic sculpture, functioning at once as an extension of the body, and as a presence which one encounters with the body."[92]

Rokeby recalls his first motivation in making this work as sharing in the "utopian hopefulness about interactivity" of the early 1980s, which he explains in terms of the promise of opening up new kinds of "transcendent" experiences:

> Interactivity would have the potential to lift us out of entrenched patterns of behavior, that would be in some manner liberating, but, importantly for me, not fantasy or escapism. . . . The early 80s was a very intoxicating time for interactivity, before the hype of the late 80s, and I, at least, had some very high hopes.[93]

In his interview for this book Rokeby further elaborated on this early mindset about interactivity, calling it "somewhat of a techno-utopian fantasy. With some weight given to it from Cage and Duchamp and other stuff. So, it wasn't purely techno-utopian but it was pretty idealistic."[94] *Very Nervous System* indeed soon came to the attention of Derrick de Kerckhove, the University of Toronto literature professor who carried forward the mantle of McLuhan's Centre for Culture and Technology as codirector after its founder's death in 1980. De Kerckhove enthusiastically promoted Rokeby's work as a paradigmatic case of new forms of media art serving as an "extension of the senses." *Very Nervous System* was displayed at art and technology festivals, art galleries, and at computer industry venues, like the annual trade and academic graphics conference, SIGGRAPH. It was in observing users at such venues that Rokeby started to become disaffected from his initial ideals for interactivity. He began to notice people moving in synchronization to the sounds generated by the system, not surprisingly, since "the system was programmed initially to respond identically to identical movements." In learning the system, in other words, the "third time becomes a command." He noticed that "people would synchronize to rhythmic sounds, without realizing they were synchronizing to it, then claim the system was very intelligent, because it had [they mistakenly thought] synchronized to them."[95]

This misapprehension disturbed Rokeby and led him to several years of scrutiny and reconsideration. Eventually he made revisions to the system and wrote an influential article theorizing the relationship between subjectivity and control in interactive art.[96] Dissatisfied with simplistic reflection of gesture as predetermined stimulus response, Rokeby would formulate his more developed thinking on interaction as being more about "encounter" than "control."

> The interactive artist must counter the video-game induced expectation that the interactor often brings to interaction. Obliqueness and irony within the transformation and the coexistence of many different variables of control within the interactive media provide for a richer, though perhaps less ego-gratifying experience.[97]

Rokeby arrived at this more nuanced conception of interaction in parallel with other voices challenging the utopian "fantasy" of empowerment through interactivity, pronouncing a critique that would grow in acuity as "user-friendly" computers became exponentially more common in the later 1980s. Rokeby spent a month as a resident artist at the Banff Centre for the Arts in Alberta, alongside American composer, performer, and theorist

George E. Lewis. The Centre had just opened a new program for media arts, and the pair, along with Canadian computer musician Daniel Scheidt, were among its first invited artists.

For Lewis, the key relationship to work out was that between interaction and improvisation. In an early formulation of his thinking, presented at an art and technology conference also at Banff in 1986, Lewis turned to precedents in electronic music built using hardware circuitry capable of generating musical "behavior" in order to distinguish between, on one hand, software-based compositions that merely trigger precomposed material at the behest of human performers, and, on the other, "intelligent instruments" that have some kind of internal agency of their own.[98] Scheidt, in turn, had been influenced by his studies with another Canadian computer musician, Martin Bartlett (1939–1993), who worked with both analog circuits and then software-based compositions with a sensitive ear tuned to maximize variety and discovery, rather than routine and predictability. Scheidt's personalized software compositions were designed as open systems for both human improvisors and generative music behaviors. Lewis would go on to publish extensively on the topic of improvisation and interactivity, initially focused on his software composition *Voyager* (1986–1988), which he characterizes as capable of generating multiple streams of music, emanating from both the computers and the humans—a "non-hierarchical, improvisational, subject-subject model of discourse, rather than a stimulus/response setup."[99]

In Rokeby's developed critique of interactivity, he was acutely aware of the rising prevalence of software applications ostensibly designed to enhance individual creativity (like graphic design and paint programs), but which in fact serve as a force of standardization and conformity. "One can look at the distribution of a creative medium in the form of a software package as a subtle form of broadcasting."[100] He expressed his approach toward distributed human-machine agency in terms of *diffusion* and employed several complementary metaphors to explain this concept. One of them compares the aesthetics of experiencing an interactive artwork to the resolution of human binocular vision into the revelation of depth: in his ideal of interactivity, Rokeby seeks to provide the user with a transformed reflection, "an image of the self from another point of view that likewise produces a sort of stereoscopic tension."[101] In another formulation, he employs the term "resonance" to indicate a similar notion of a gap or interval between two separate, negotiating agents (much as McLuhan had spoken of Canada's condition as a borderline state in terms of resonant intervals). As noted by digital art theorist Katya Kwastek, *Very Nervous System* is, paradoxically, "programmed to resist conscious control."[102] Just as George Lewis considers his *Voyager* to be a case of "techno-mediated animism,"[103] *Very Nervous System* allows users to "attune their movements to the sensitivities of the [system], in way that accents collaboration, exploration, and improvisation."[104]

Rokeby's critique and practice of interactivity in art, complemented by Lewis's parallel theorization of an improvisatory approach to interactive systems, finds contemporary expression in recent reflections by Alan Kay, one of the initial designers of the personal computer, who had hosted Ron Baecker at Xerox PARC in the 1970s. Castigating the way computer interfaces have dumbed down users by emphasizing "ease of use" and "transparency," Kay condemns "the desire of a consumer society to have no learning curves. . . . We can contrast this with technologies that do have learning curves, but pay off well and allow users to become experts (for example, musical instruments, writing, bicycles)."[105]

Media theorist Wendy Hui-Kyong Chun sharpens and extends the kind of dissatisfaction Kay expressed about the oversimplifications of interface transparency, in ways that chime with Rokeby's own critical methodology, when she suggests that "the interface is 'haunted' by processes hidden by our seemingly transparent Graphical User Interfaces that make us even more vulnerable online, from malicious 'backdoors' to mundane data gathering systems."[106] It is not just that configured users are falsely empowered, as Kay would have it, but they are blithely unaware of the risks they are undergoing in permitting their every motion to be captured, mined, targeted, and monetized. Chun proposes instead that what is needed is to "make our interfaces more productively spectral—by reworking rather than simply shunning the usual modes of 'user empowerment.'"[107] It is well to recall that some of the earliest encounters with calculated motion at the interface, as recounted in this chapter, were in fact well attuned to Chun's admonition, in their pursuit of expressive discovery and bodily expansion.

5 Tuning, Tundra, and Treasure: Canada's Expanded Musicality

On June 1, 1964, Glenn Gould, a freshly liberated thirty-two-year-old "concert dropout," gave the commencement address at the graduating ceremony of the Faculty of Music at the University of Toronto, on the occasion of his being awarded an honorary doctorate. Having given his last piano recital in Los Angeles in April, he took the occasion to offer an "argument for music in the electronic age." Far from being a mere restatement of the musician's well-publicized preference for the recording studio over the concert hall, Gould's argument was subtle and prescient. At issue in a musical age suffused by electronic technology was nothing less than "the structure of participation" between the composer, performer, and listener.[1] The respective functions of these previously discrete agents would begin to overlap with each other, he explained, so that the listener would become more and more directly involved, participating in new creative ways. For Gould in 1964, a profound consequence of this changing relationship between activated listeners and professional interpreters and creators would be an increase in individual agency: rather than being one in a crowd subject to the forward sweep of "historic progressivism," or what Gould would later call the "tyrannical force of the *Zeitgeist*," the individual "to a large extent [would be] sheltered . . . against the buffets and fortunes of the world of taste and fashion." Indeed, the "ultimate gift of electronic culture to art" would be nothing less than "a total rethinking of the nature of individuality."[2]

Over the next year, Gould would develop these ideas in conversation with Marshall McLuhan, collecting their combined thoughts in a ninety-minute broadcast over the CBC in 1965.[3] This program would become the intellectual basis for Gould's most famous essay, "The Prospects of Recording," which appeared in *High Fidelity* the same year.[4] In the broadcast dialogue,

McLuhan quoted a phrase from James Joyce's *Finnegans Wake,* "My consumers, are they not my producers?" Audiences will become more creative, he suggested, like composers; musical participants will become "programmers," and access to music will "become encyclopedic" rather than being limited to today's hit parade, so that "musics will become plural."[5] For his part, Gould explained the reconfiguration of the relationship between actors in the musical activity in greater analytic detail than McLuhan did, considering the technological basis for these changed relationships. To use terms I have employed previously in this book, Gould forecast a future for electronically mediated music as both *object* and *process.* Foreshadowing what Alex Galloway argued almost fifty years later, Gould already thought of the prospects of recording in ways akin to what Galloway calls a "fertile nexus," comprising various "active thresholds"[6] between listener, performer, and composer. Gould extrapolated forward from the controls then available to consumers in the high-end audiophile market, like variable-speed turntables, filters for control over frequency ranges, and tape splicing on reel-to-reel tape recorders, envisioning a "new listener" who had at least some of the prerogatives of composers and conductors. "Dial twiddling is in its limited way an interpretative act," Gould observed, and while these controls were but "primitive regulatory devices," they would evolve greatly once techniques available in electronic music laboratories became available to consumers in home equipment.[7]

Gould was intimately familiar with the state of the art in avant-garde electronic music composition and had closely observed the work going on at the University of Toronto's Electronic Music Studios. During those same years in the mid-1960s, UTEMS had become one of the most advanced studios in the world, equipped with Hugh Le Caine's latest instrumentation, and supported experimental production by a string of visiting composers. Pauline Oliveros, to give one example, made some of her acknowledged masterpieces in tape music composition at UTEMS, later recalling Le Caine's warmth and generosity to her personally, and the extraordinary excitement she felt using his 20-channel loop machine, or Special Purpose Tape Recorder.[8] In an article entitled "Malleability and Machines: Glenn Gould and the Technological Self," Canadian historian of technology Edward Jones-Imhotep reconstructs in detail the musical possibilities at UTEMS that informed Gould's outlook directly.[9] Jones-Imhotep zeroes in on why Gould, the renowned performer of complex polyphonic works by composers such as Bach and Schoenberg, so appreciated UTEMS's various instruments that favored "questions of counterpoint." Inventions like Le Caine's multichannel loop machine allowed for "the synchronization of melodic units into a contrapuntal texture" and "the organization of sounds in extremely complex rhythmic continuities,"[10] both of which were difficult to do with the knob-controlled oscillators, one- or two-channel tape recorders, and monophonic analog synthesizers available to composers in other labs around the world at the time. Therefore, when Gould speculated in his article on "Prospects of Recording" about new "participational possibilities,"[11] he was doing so with a concrete grasp of actual instruments that he imagined might eventually be converted into consumer-level devices with convivial interfaces. Gould's essay concludes with another forecast that is sometimes misconstrued by modern readers:

> In the best of all possible worlds, art would be unnecessary. Its offer of restorative, placative therapy would go begging a patient. The professional specialization involved in its making would be presumption. . . . The audience would be the artist and their life would be art.[12]

Here Gould speaks, and not for the last time, as a technological moralist, changing the terms for how to think about the role played by art in everyday life. He is not making an argument about the actual end of art as such,[13] but rather an end to the way it had come to be commodified and served up to consumers as a placative or escapist "therapy." Gould was already starting to construe a participatory, technological musicality in quasi-spiritual terms, a tendency that I will return to later in my discussion of his "contrapuntal radio" compositions.

Gould's address to the University of Toronto graduates in 1964, then, articulates a vision for music in the electronic age that can now be seen in parallel to Norman McLaren's proto-computational films of the 1960s, as discussed in chapter 2. The prospects Gould imagined for recording anticipate some of the salient features of the digital era to come, from interactive interfaces to individualized network music delivery systems. As well, the original works he composed for radio elicited extensive commentary by authors in fields outside music, much as McLaren's films had "turned on" the electrical engineers in the early computer graphics culture. The parallels between these two iconic Canadian cultural figures are suggestive, but they only go so far. Indeed, the differences between music and cinema with respect to their relative susceptibility to, and affordances for, digitalization are fundamental in drawing out the implications of Gould's prospective vision more broadly.

While McLaren's proto-computational sensibility was technically grounded in the way the images in his films were given a notational structure amenable to programming, music in the Euro-American tradition that formed the primary basis for Gould's thinking had already been organized notationally for 1,000 years. The primacy of this symbolic order of musical representation, based in the medium of print, was weakening as recordings came to the fore after 1900, and also as experimental composers started from around 1950 to devise alternative schemes to describe more open kinds of structure. Both these attenuations of music notation's prescriptive function occurred well in advance of Gould's prognostications in the mid-1960s. But the notational *affordances* of Western music, as anthropologist Mary Douglas notes, are but a special case of a general form of symbolization that can spark, not inhibit, innovation: notation "define[s] the units of the world that are taken for exemplars" in the categories to which it is applied, and this categorical thinking can "enhanc[e] collaboration to reach a specified standard."[14] Notationality construed thus, as a general device for coordinating between various kinds of objects and actors, did not dissolve in music with the direct inscription of sound in recording or the experimental explorations of new music. Rather, it spread out to provide a model for the ways in which music might be aligned articulately with other media, in particular electronic images, which had already seen a long history in "visual music,"[15] but also the recorded spoken word. Gould's own "contrapuntal radio" was in just this way based on a treatment of language as quasi-musical motivic units, then subjected to compositional operations. Just as, in Gould's account, the boundaries between listener, composer, and performer were becoming less rigid, then, the longer history of musical notation reveals that the lines separating music from other art forms were also becoming weaker, more like "active thresholds." This had the effect of opening music up further to various kinds of intermedial reconfiguration. This *expanded* field for musical practice and thought paralleled developments in cinema during the same years, as indicated by an influential book by Gene Youngblood that chronicled cinema's transition into the analog electronic and digital eras.[16]

This opening out or expansion of music to new or renewed relationships with other art forms occurred on a broader scale as well, in a way that was particularly salient in Canada during the period from the 1960s on. In prior scholarship on Canadian music and communication covering this period, an emphasis has been placed on the hypothesis, now typically associated with McLuhan, of a retrieval in the modern era of an older paradigm of primarily auditory awareness. This hypothesis, while remaining somewhat controversial to this day,[17] was influentially formulated through the concept of "acoustic space" within the interdisciplinary hotbed of the Explorations group at the University of Toronto in 1953.[18] Psychologist Carl Williams had been the first to give a specific account of acoustic space in terms similar to those that are so familiar today, mainly because of how germinal it later became in McLuhan's writings.[19] As Williams initially wrote, "the essential feature of sound . . . is not that it be located at a point, but that it *be*, that it fill auditory space. . . . Auditory space has no boundaries in the visual sense. . . . Auditory space lacks the precision of visual orientation."[20] Within Canada, the concept of acoustic space and the implications of reconfiguring research and creative priorities around sound was taken up with particular energy and creativity, most pointedly in the research initiated by composer R. Murray Schafer on the "world soundscape" in the early 1970s. The goal of this project was to study "the changing soundscape and to determine how these changes might affect people's thinking and social activities."[21] Beyond just recording, cataloging, and analyzing changing soundscapes, the constructive goal of the project was to create a new interdisciplinary design discipline oriented ecologically around sound, a goal that within decades would find adherents worldwide.

In this chapter I will argue that between the 1960s and 1990s this expanded musical field was one of particularly fertile creativity in Canada. This history of Canada's "expanded musicality" begins in Vancouver, the birthplace of the World Soundscape Project, and then proceeds across the country in a consideration of various scenes for improvisatory music that appeared during this period, all of which foreground the innovative use of media technologies. Zooming out from this historical context, I attempt to outline a group of "electroacoustic convergences" that accompanied and occasionally intersected with these emergent scenes, showing how Canadian electronic music over this span of time came to foreground interrelationality with other art-forms, in both a practical and theoretical sense. Returning in conclusion to Gould's own creative work in contrapuntal radio during the 1960s and 1970s, I offer a close reading of these broadcasts, looking both to the way they combined music and speech at a granular level, and for the broader questions it raises about the place of humanity in a world that was beginning to be understood, during these years, as being enveloped within a network of encompassing technological systems.

The World Soundscape Project and the Intellectual Formation of Acoustic Ecology

When composer R. Murray Schafer (1933–2021) addressed the Royal Society of Canada in 1967 on the future of the arts, he told them that "it is not the arts themselves but the *relationships between the arts* that will characterize most of the creative thinking of the immediate future."[22] At the time he made this declaration, he had served for two years as one of the charter faculty members of a radically experimental Centre for Communication and the Arts at Simon Fraser University.[23] As seen in my discussion of dance technology in chapter 4, SFU was founded on a mandate to support interdisciplinarity, and this ideal would flourish in a highly productive way in the grouping of contemporary arts, communication, and computational sciences within the dedicated Faculty for Interdisciplinary Studies (IDS). It was in fact within the Department of Communication Studies of IDS that Schafer founded the World Soundscape Project (WSP), a department that grouped scholars from many diverse fields across the social sciences. This communicational context was important for the success of the WSP, and for soundscape studies over the long term, as one of the project's explicit aims was to move beyond music as such, to become what was variously called acoustic design or acoustic ecology.

While Schafer occupied his academic post at SFU from 1965 to 1975, he sought to balance his creative work as a composer with both scholarly research and pedagogical innovation. In each area, he enjoyed widespread recognition, but he is often better known for only one or at best two of these contributions. As a composer he is recognized for a body of work ranging from instrumental and choral music to environmental operas, an oeuvre displaying the unusual amalgamation of twentieth-century experimental techniques (including graphic notation), twelve-tone methods, and a romantic expressive sensibility. In some countries, like Brazil and Japan, he is best known as a radical music educator. In contemporary sound studies, he is acknowledged as one of the field's founding figures.[24] When he founded the WSP he combined all three. As previously noted, from the outset the WSP's formal objectives were both synthetic and creative: "to bring together research on the scientific, sociological, and aesthetic aspects of the sonic environment."[25] While in critical terms the WSP sharply attacked both the rising levels of industrial-caused "noise pollution," as it was called, and the increasing encroachment of background music into public space, Schafer meant the WSP to take a constructive, design-oriented approach, not just a critical one. In the *Handbook for Acoustic Ecology*, edited by Schafer's close collaborator Barry Truax and published in 1978, this positive aim was expressed thus:

> We regard the sounds of the environment as a great macro-cultural composition, of which man and nature are the composer/performers. . . . People must be stimulated to take a more *active* part in their acoustic environment. . . . The [World Soundscape Project] combines a musical "common sense" with many functional activities, including architecture, psychology, acoustic engineering, urban geography, history, communications, and many other disciplines. . . . [The aim is] to lay the foundation for a new field of studies which might be called Acoustic Design or even Soundscape Ecology. Our aim is to provide coherent facts by which decisions can be made not only to control, but also to compose, the acoustic environments of the future.[26]

Schafer's own synthesis of the project appeared in 1977 as *The Tuning of the World*, a book offering mainly descriptive evidence of changing sonic environments around the world and across history. It was also a passionate appeal to composers to take on a design role in the creation of more humane acoustic environments, which started by learning anew "how to listen," a task he called "ear cleaning."[27] Furthermore, to reeducate the senses, one had also to study their individual relationships in detail. For inspiration in this, Schafer turned to the Bauhaus of the 1920s, where, he argued, "the most important revolution in aesthetic education in the twentieth century [was] accomplished."[28] The Bauhaus teaching that most impressed him was explicitly relational, that is, in the way it brought together the fine arts (e.g., the work of Klee and Kandinsky) with the industrial crafts. Similarly, he argued,

> it devolves on us now to invent a subject which we might call acoustic design, an inter-discipline in which musicians, acousticians, psychologists, sociologists and others would study the world soundscape together in order to make intelligent recommendations for its improvement.[29]

The definition of soundscape in the 1978 *Handbook* is "an environment of sound (sonic environment) with emphasis on the way it is perceived and understood by the individual, or by society. It thus depends on the relationship between the individual and any such environment."[30] One of the enduring techniques emerging from the WSP was the soundwalk, urged as a "form of active participation in the Soundscape . . . to encourage the participant to listen discriminatively, and moreover, to make critical judgements about the sounds heard and their contribution to the balance or imbalance of the sonic environment."[31] As Gould had emphasized the agency of the new listener when mediated by technology, the WSP stressed the direct participation of the listener in the environment through activities such as soundwalking, taken both as educational and aesthetic practice. But Schafer's own creative output was to remain focused on instrumental and vocal composition within, and increasingly outside, the confines of concert halls.

After giving up his Simon Fraser professorship in 1975 to focus on composition, one of the first works Schafer completed was the prize-winning String Quartet no. 2 (*Waves*). Schafer structured this work around recorded measurements made by the WSP of ocean waves; the period of these waves set the phrasing for an undulating rhythmic texture. Schafer sought to give the quartet a "liquid quality in which everything is constantly dissolving and flowing into everything else. . . . The material is not fixed but is perpetually changing, and even though certain motivic figures are used repeatedly, they undergo continual dynamic, rhythmic, and tempo variations."[32] To specify this liquid quality Schafer gave the players the interpretive flexibility to morph their figures in sync with the fluctuating period of the waves. In this way, the quantitative measurements of the soundscape became the model for a musical work with a strikingly natural qualitative flow and sense of spontaneity: the natural context thereby informed aesthetics in a complex variegated way.

Several years later, Schafer took the contextual direction further by creating the first of what would become the *Patria* cycle, a set of ten large-scale environmentally situated works. *The Princess of the Stars*, which in 1985 received a widely attended series of performances at the shores of a mountain lake in Banff National Park, began before dawn with the audience being led by costumed ushers to the lake's shore. A plaintive soprano aria was heard reverberating across the surrounding mountain peaks. Woodwind, brass, and percussion instruments were played by musicians scattered around the lake shore. Oversized puppets of birds and animals mounted on canoes were animated by characters voiced by sound poets, in a narrative culminating in a sea battle. As the sun itself rose, a shimmering figure of the sun mounted on a canoe appeared to end the conflict. Schafer considered this ritual opera to be his most "Canadian" creative work.[33] It serves as the prologue within the *Patria* cycle; the last of the cycle, *Wolf Shall Inherit the Moon*, takes several weeks to perform.[34]

Serving as Schafer's assistant for nine months during the first years of the WSP, German-born Canadian composer Hildegard Westerkamp (b. 1946) is the figure most responsible for developing soundwalking into an acknowledged field of educational as well as aesthetic practice.[35] In a 1974 article, Westerkamp defined what would become the consistent theme in her writings on acoustic ecology, to establish "balance and focus" in both listening and sound-making. By balance, Westerkamp meant an equilibrium between the sounds one makes oneself and the sounds one hears.

> Try to move
> Without making any sound.
> Is it possible?
>
> Which is
> the quietest sound of your body?
>
> If, however, you cannot hear the sounds you yourself produce,
> you experience a soundscape out of balance. Human proportions have
> no meaning here. Not only are your voice and footsteps inaudible but also
> your ear is dealing with an overload of sound.[36]

By "focus," Westerkamp refers to the ear's selectivity and ability to change levels of attention as well as awareness. By increasing the degree and intensity of creativity accorded to the listener, in effect expanding the aperture of sonic reception, Westerkamp is elaborating on an enlarged "structure of participation" in music in a way consistent with that articulated earlier by Gould. Emphasizing depth and engagement in the act of listening, Westerkamp's work was also enriched by similar orientations in the work of American composer Pauline Oliveros, with whom she studied in the workshops Oliveros offered during her visits to Vancouver in the late 1970s. Oliveros's practice of deep listening also entails a comprehensive capacity to tune and balance ranges of attention and awareness, and grew out of a sonic meditative practice.[37] Westerkamp too combined music composition with exercises in aural awareness, and created some of the earliest works in the genre of "soundscape composition."[38] As both a composer and an environmental activist, Westerkamp saw herself carrying forward Schafer's initial critical engagements from the 1970s, alongside Oliveros's equally powerful spiritual influence. Creatively, Westerkamp considered environmental sounds to be "her instruments." This ecological orientation led her, in 1985, to write that she was

> no longer interested in making music in the conventional sense. I am interested in addressing cultural and social concerns in the musical idiom. I want to find "voices" of a place or a situation that can speak most powerfully about a place/situation and about our experience in and with it. I consider myself an ecologist of sound. . . . To think about acoustic ecology is also important to me as a woman. What are women's soundmaking capacities/powers/opportunities in the face of overload of sound and music input?[39]

With Barry Truax, Westerkamp was instrumental in helping found the World Forum for Acoustic Ecology in 1993, an organization that aims to formalize the field through regular conferences and the publication of a scholarly journal.[40] The World Forum was the outcome of an international conference, The Tuning of the World, that marked Schafer's sixtieth birthday and brought together a wide international audience. A composer and computer technologist, Truax (b. 1947) took over Schafer's position within the SFU Department of Communication Studies, and become the main theoretician of the field, in a series of articles and books (starting with his contribution to the aforementioned 1978 *Handbook*) continuing to the present in which he has recently put forward a cogent formulation of "context-based" musical composition.[41] In his treatise on acoustic communication, Truax defined the listener's role in a three-way mediating relationship with sound and the environment,[42] and he has insisted that "real world contexts inform the design and composition of aurally based work at every level."[43] This includes music's placement within sociocultural contexts as well as those of the natural environment. Eschewing the tendency in Euro-American classical music to reward absolute music and "abstracted" listening, Truax has emphasized the crucial importance of the knowledge that listeners bring to bear on what they hear, an approach which, he explained, "differs fundamentally from [one] that utilizes sounds related only to each other in an apparently autonomous form."[44]

Until his retirement from SFU in 2015, Truax bridged the academic worlds of music and communication, but he also is an important software developer and expert in computer music.[45] In this way, he may be considered Canada's "learned musician," using the phrase taken from the title of

Christoph Wolff's biography of Johann Sebastian Bach.[46] The comparison is made here in a formal, not evaluative sense, drawing on Wolff's characterization of Bach's combination of "scholarly thought and theoretical command" applied to the "shaping of musical material."[47] Another point of comparison is that Bach's music was directed toward and contextually grasped within a particular social environment—the religious life of Lutheran Germany—and Truax's music too is intended as a contextual contribution, in a sense therapeutic, toward what he himself calls the "re-integration of the listener with the environment in a balanced ecological relationship."[48]

Truax's electroacoustic composition *Riverrun* was produced in 1986 using his own software system for real-time granular synthesis to simulate another kind of natural scene, a river soundscape. Granular synthesis is a method which uses thousands of microsonic particles called grains to provide the building blocks for composition; for Truax, achieving live control of such microsonic flows was essential to enable the freedom of "real-time composing." This freedom entails an abandonment of "linear modes of compositional thinking, which result in deterministic output (e.g. score or sequencer driven)," and instead replaces it with "process-oriented multi-task strategies for real-time execution."[49] For music theorist Mara Helmuth, "listening to *Riverrun* . . . more closely connects the listener to the physical world through a metaphor of water sounds, creating an experience related to that of a natural wonder. While the work was created with real-time synthesis, it is heard as a river soundscape."[50] Much like Schafer's *Waves*, Truax's *Riverrun* is a work that combines "stasis and flux, solidity and movement."[51] These coexist, as Truax notes, "in a dynamic balance similar to a river . . . much like environmental sound generally and water sound in particular."[52] With this work, which was awarded the Prix Magisterium in electroacoustic music at Bourges in 1991, Truax knew he was "breaking into a new domain" owing to the emotional intensity produced by his program's capacity to generate such an immense number of events in real-time.

> It felt like such an incredible, powerful force that I was unleashing, even on myself. I never played it in public for quite some number of months. And I never even let anyone hear it in the studio, because of the sheer power of it. It's like, "My God, what have I created here?"[53]

The quasi-autonomous power Truax felt emanating from his synthesis program was in fact balanced by his own ability to shape its outcomes, to "compose in real-time."[54]

Taken as a whole, the contributions of Truax, Westerkamp, and Schafer were central to the establishment of acoustic ecology as an expanded mode of musicality in Canada and far beyond. Each in their way brings to the fore the creative importance of listening as an active pursuit, with varying degrees of engagement with new media. All were aligned with the ethos of "acoustic space" espoused by McLuhan, who was enthralled with Schafer's work on the *Tuning of the World,* taking it as the accurate and articulate fulfilment of his own more inchoate intellectual formulation of a "sonic imaginary."[55] While Schafer only made electronic music early in his career, Westerkamp's and Truax's uses of electronic media were groundbreaking and highly influential, both aesthetically and technologically. Westerkamp's contextually grounded soundwalks in fact grew out of a series of community radio broadcasts of soundscapes, only later becoming active forms of guided environmental exploration.[56] And Truax's fluent combination of improvisatory expression with computational programming expertise, including systematic advances as well as compositional invention, offers another exemplar of the alternative technological ethos seen in multiple cases already in this book. That is, the capacity to maintain a high degree of gestural freedom is enhanced, not hindered, within a highly elaborated algorithmic musical space.

Improvisatory Scenes, Local Differences, and the Special Case of *Musique Actuelle*

"Composition in real time" is, by some accounts, effectively interchangeable with the word "improvisation," and during the 1970s and 1980s, Canada was replete with local scenes for improvised music making. In Vancouver, during the same years as the World Soundscape Project was carrying out its groundbreaking research, composer Martin Bartlett developed electroacoustic methods that aimed to have a degree of self-generating agency. Bartlett began to explore generative processes in music production as a graduate student at Mills College in the San Francisco Bay area.[57] There he began to experiment with primitive reactive systems that possessed a degree of autonomous behavior. In a 1978 article in the *Computer Music Journal,* Bartlett explained that his goal was to give the computer "its own personality and sound," and to do so in such a way that small adjustments of the set-up could produce sonic results "comparable to the way an Indian musician might exploit the pitch and duration associations of a particular raga or tala. The instrument is in a continuous state of change, with modules being retired and new ones constructed as my musical ideas and performance aims change." Importantly to Bartlett, even without any input, this digitally controlled analog system was meant to be able to generate "complex event sequences" with musically interesting results, "having that peculiar combination of volition and purposelessness which seems so characteristic of computers when they are feeling playful."[58] Bartlett was a bedrock of the Vancouver experimental arts world, having been one of the founders of the artist-run center the Western Front; he also taught a rising generation of composers to program interactive music systems with a degree of "volition" that can only be called improvisatory. A 2017 film, *Electro-Pythagoras*, directed by artist Luke Fowler, celebrates Bartlett's life and work, explaining in detail the basis for his insistence on "intimate relationships with technology" using handcrafted electronics. For Bartlett, this artisanal direct control over design and behavior "soften[s] the technological anonymity, creating individual differences [and] making each instrument a topography of uncertainties with which we become acquainted through practice."[59] Bartlett also initiated an intercultural workshop series at Simon Fraser University in the late 1980s combining interactive computer music with Indonesian gamelan studies.[60]

One of the instructors in these SFU workshops was Daniel Scheidt (previously introduced in chapter 4), who brought formal training in computer science to the challenge of creating "interactive software *instruments*," treating the computer in performance as one would any traditional instrument. The individuality of different pieces is defined by the properties given to each instrument; that is, "compositional algorithms are embedded in the software which allow each instrument to have *knowledge* about the music it is intended to play," which, for Scheidt, was crucial to enable performers to "interact with compositional elements of the piece rather than simply following a fixed score."[61]

Toronto too saw a flourishing of improvisatory electronic music, notably during the tenure of US composer David Rosenboom at York University's experimentally inclined music department from 1970 to 1980. An influential teacher as well as virtuoso instrumental performer, Rosenboom designed brainwave-controlled interfaces that helped instigate a short-lived surge of interest in biofeedback across the arts during the 1970s.[62] One of

Rosenboom's signature works is *On Being Invisible* (1976–1977), which he characterized as an "attention dependent sonic environment."[63] Rosenboom's reactive music system monitors electrical activities within the brain to extract patterns that are converted to sound using banks of resonant filters.[64] The concern articulated throughout this chapter for musical processes mediated by responsive technological systems is made explicit by Rosenboom:

> It is essential . . . that the performer constantly ride a border line between being, on the one hand, an initiator of action and, on the other, submerging him/herself in processes larger than him/herself. This requires that the performer become adept at manipulating his/her state of consciousness, application of willful actions, and the energizing of programmed personal response modes.[65]

Also in Toronto, a downtown improvisation scene flourished in close connection to the visual arts, including Michael Snow who cofounded the Canadian Creative Music Collective along with other painters, often using synthesizers in their performances. The Canadian Electronic Ensemble is one of the earliest and longest-lived groups dedicated to live electronic music. Elsewhere in Canada, at the Banff Centre, in Alberta, an international workshop on "improvisational composition" was held in the winter of 1980 which combined classical, avant-garde, jazz, and electronic musicians in an intensive program whose spinoffs were registered in various Canadian centers and abroad.[66] From the mid-1980s, a unique Maritime approach toward bridging experimental music with improvisation and environmental concerns occurred in St. Johns, Newfoundland, during the annual Sound Symposium.[67]

In Québec, another unique confluence of radical experimentation in music and electronics occurred, one that was inflected strongly by the province's political and social transformation in the 1960s. Québec cultural life was undergoing rapid mutations within the Quiet Revolution after 1960, which brought on a rapid diminishment in clerical control over education and the arts, as well as the equally rapid rise of popular support for the political movement advocating separation of the province from the Canadian federation. For University of Moncton scholar Anne Robineau, the phrase "cultural laboratory" applies well to the entire social context surrounding the Quiet Revolution, and an important element in the expression of emancipation and freedom was found in the "free jazz" ensembles established in Montréal during the 1960s.[68] This improvisation-inflected music scene, quick to embrace electric and digital technologies from popular sources as well, also vigorously opposed the orthodox compositional style known as postserialism, which had powerful academic supporters in Montréal. The phenomenon that would come to be called *musique actuelle* was born in opposition to the "intellectualism of contemporary music and the rigidity of written (notated) music."[69] If "music had an enemy," according to Philippe Ménard, one of the members of the 1970s Montréal musical underground, it was "the score": equating written music with death and improvisation with life, Ménard also cites with approval the ecological rhetoric of Murray Schafer, but goes further in demanding that music must have the function of "denunciation." In a nod to Adorno, Ménard writes that art can denounce "social reality," break with it, and serve as a great Refusal (*"être le grand Refus"*).[70]

Jean Derome, a multi-instrumentalist and core member of the Montréal *musique actuelle* scene, is also the author of several manifestos that define its broad aesthetic, writing that it crosses between but has no specific allegiances to jazz, the "classical avant-garde," or European improvised music, but also reaches out to include blues, traditional fiddle tunes, bluegrass, and the folk music of the deep Québec countryside: "Musique actuelle veut dire musique métisse."[71]

In 1980, an independence referendum was held in Québec, the first of two that would come close, in the mid-1990s, to achieving sovereignty for the mainly francophone province. The energy building up to this first election had received a powerful boost by the visit of French President Charles de Gaulle in 1967, during which he proclaimed support for a "free" Québec, and many moderates were radicalized by the Trudeau's government's draconian crackdown in 1970 on a terrorist insurrection led by the Front de Libération du Québec (FLQ). All these and many more sociohistorical forces came together in an extraordinary way when composer and guitarist René Lussier was commissioned by the French service of the Canadian Broadcasting Corporation, Radio Canada, to create *Le trésor de la langue*, a radiophonic program that he considered an "intuitive sociological portrait" of Québec.[72] A close collaborator of Derome and stalwart of the *musique actuelle* scene, Lussier recorded the voices of everyday Québécois from across the province and collected archival tapes covering the charged events of its recent past. *Le trésor* was broadcast in 1989, while another movement toward independence was taking off. It won an international prize for best French-language public broadcasting program in that year. It also was nearly suppressed by the top management of Radio Canada, because some of the material Lussier used, like the hostage demands of the 1970 FLQ, were seen to be so incendiary.[73]

Stylistically, Lussier's work is based on a meticulous musical transcription of the spoken word, arranged for an ensemble of 12 musicians, including Jean Derome and Fred Frith. These transcriptions were played by instrumentalists, either alongside the initial voices, or as highly virtuosic instrumental solos "voicing" the exact contour and cadence of the texts. As an intermedial music-text composition, *Le trésor* is not dissimilar to Steve Reich's string quartet with sampled voices, *Different Trains* (1988), in the way texts are accompanied by instrumental performances exactly mimicking the contours of their pitch, rhythm, and inflection of words. Unlike Reich, Lussier's stylistic range is eclectic, and the pluralist influences of *musique actuelle* inform this immense compositional variety, sometimes with abrupt juxtapositions and rapid-fire changes. In addition to de Gaulle's "Vive le Québec Libre" speech and the FLQ manifesto, Lussier included the concession address given by independence leader René Lévesque after the 1980 referendum defeat, a slow blues with low-pitch glissandi expressing the moans and disappointments of that moment of collective trauma to a large portion of the Québec population. The voice of union leader Michel Chartrand is heard giving a fiery anticapitalist, pro-independence speech, which Lussier rhythmetizes over a driving groove, punctuated by the words "c'est ça qu'on va faire" (that's what we're going to do).

The fixed-media version of *Le trésor* was recorded and produced by Lussier himself, with special permission granted in order to avoid the union restrictions governing Radio Canada sound technicians. Working with American composer and electronic musician Bob Ostertag, Lussier later devised a live version in which the vocal parts could be performed using keyboard-controlled digital samplers. This version was performed in festivals in North America and Europe, and became the subject of a 1996 documentary film, *Le trésor archangel.*

The expanded musicality of Lussier and his collaborators can be located within the close linkages and alignments they realize between the phonological level of minutely analyzed spoken language and their microtonal and microtiming musical performances and improvisations. Here the "new listener" forecast by Gould and widely developed in the acoustic ecology group is less at the fore, in terms of Gould's restructuring of musical participation, while the range of the composer expands to embrace the construction of text-sound hybrids not only within fixed media (as seen already at the UTEMS, or in Gould's own contrapuntal radio works) but also in live performance with virtuosic improvisors. In another sense, however, the listener's role is also expanded in the way the voices of the Québec people themselves are made palpable and direct in this sonic portrait of a nation.

Perhaps most affecting of all the movements in this hour-length work, and forward-looking in the light of later developments in Canada's treatment of indigenous peoples, Lussier composed a movement based on a poem by Richard Desjardins entitled "Qui se souvient?" (Who remembers?)

When Desjardins's voice is heard after a searing guitar introduction laden with distortion and a shrieking saxophone improvisation, it is not instrumentally embellished but presented unadorned, unlike the intricate instrumental elaboration of the voices of de Gaulle, Lévesque, Chartrand, and the FLQ spokesman. The final words of the poem, "Go ask the Iroquois . . . while they're still around," are not accompanied by any instruments at all.

Electroacoustic Convergences

So far in this chapter, I have emphasized music in an expanded field, widening its set of connections with social, natural, and intellectual contexts, as well as intermedial linkages with other art forms or materials. This expanded musicality was happening globally after 1950 but appears to have been particularly fecund in Canada. Québec offers an example of how ideas originating from abroad were adopted and modified in innovative ways. *Musique concrète* is a genre of experimental music composition based on the editing and mixing of recorded sounds, and emerged from the research undertaken in Paris after 1945 by Pierre Schaeffer. Schaeffer, an engineer by training, was also a prolific theorist and pedagogue, who promoted a phenomenological stance toward recorded sound that emphasized its semantic autonomy by stripping away anecdotal references to its original source. The term *acousmatique* signifies this reduction of sound from its history or meaning, enabling it to become, in the words of Francis Dhomont, one of its French practitioners who was also active in Québec during the 1980s, an "autonomous sonoric art, in the same way that cinema, which is not theatre, remains, like theater, a visual art."[74] But the reception of Schaefferian *musique concrète* in

Québec was from the outset more varied and pluralistic than indicated by Dhomont's purist emphasis on sonic autonomy.

Québecoise composer Marcelle Deschênes (b. 1939) trained under Schaeffer in Paris for four years, starting in 1968. Upon her return to Québec she became an influential educator who, according to a recent study by Patrick Valiquet examining her own personal archives, extended Schaefferian doctrines in original ways.[75] Influenced in part by Paul Klee's Bauhaus-era teachings on correspondences between drawing and music, Deschênes's approaches went beyond Schaeffer's exercises for listening and analysis, to (in the words of Valiquet) "task her students with the invention of new graphic and theatrical strategies emphasizing what she referred to as the 'dialectical' entanglement of oral and written musical traditions."[76] In her own creative work, she frequently collaborated with visual and theatrical artists, and over her thirty-five-year university teaching career exposed younger artists to these kinds of intermedial approaches. Composer Jean Piché (b. 1951), for instance, who was a student of Deschênes at Laval University (as well as of Truax at SFU), developed a distinctive approach to *vidéo-musique* that explicitly avoids simple associative techniques, while aiming to "transform an otherwise detached, and essentially acousmatic experience into a collective multimediatic one."[77] Piché applies techniques from the electroacoustic studio, like granulation and spatialization, to video composition, to produce what he puts forward as a "new form of mediatic expression."[78] In contrast, Robert Normandeau (b. 1955), another student of Deschênes, has composed a body of purely acousmatic works which nonetheless were theorized as a "cinema for the ear." Precisely "in order to be autonomous," Normandeau claimed, acousmatic music needed to follow the model of cinema, and he wrote a doctoral thesis to establish the parallels with film techniques that he drew on in his works.[79]

A different kind of electroacoustic convergence, this time implicating music within still-controversial debates on intellectual property, appeared in the 1980s when Toronto composer John Oswald (b. 1953) coined the term "plunderphonics" to designate his practice of "audio piracy as a compositional prerogative."[80] Oswald explicitly aligned himself early on with the Gouldian concept of expanded listening, as when he wrote that his "listening system has a mixer instead of a receiver, an infinitely variable speed turntable, filters, reverse capability, and a pair of ears. An active listener might speed up a piece of music in order to more clearly perceive its macrostructure, or slow it down to hear articulation and detail more precisely. Portions of pieces are juxtaposed for comparison or played simultaneously."[81] These were the basic operations he used in his genre-defining plunderphonic compositions, the most notorious of which is *DAB*, based on Michael Jackson's hit single, *Bad*, which triggered a legal crackdown against the practice by the Canadian music industry in 1990. *DAB* uses a limited palette consisting just of tape razor edits and digital sampling of the original. It excluded any manipulation of timbre, of pitch, or added reverberation. "What is left is manipulation in time. Fragments of the original music, varying in duration from about 25 milliseconds to phonemes and sonemes, to partial song phrases, have been superimposed and re-juxtaposed."[82]

In a detailed analysis of the opening section of *DAB*, Oswald explains how two kinds of bass line permutation were produced completely separately from each other, then "thrown together in the mix. This lack of synchronization was intentional, reflected the prime thesis of DAB, which is *relationary* in intent." With the word "relationary," Oswald is referring to his effort to disturb the mechanical precision of quantizing beat box machines, and also to make the resulting tracks "independent within the scale of rhythmic nuance":

> "DAB" mediates between a general similarity to its source material and abstraction. It also mediates between pop/rock predictability, the rhythmic idiosyncrasy available to improvisors, and the timbral mediations found in the electroacoustic genre. . . . Perfection has been messed with by improvisation. Very popular music meets its extremities.[83]

For sound art historian Douglas Kahn, Oswald's plunderphonic pieces offer a "complex weaving of conceptual and affective references to musical cultures, intellectual property issues, technological repetition."[84] Microediting of new sonic realities is again at the foreground, as with Gould and Lussier, but now with an explicit and provocative challenge to the paradoxes of copyright conventions as they were starting to collide with new technologies in the quickly arriving digital age.

In Montréal, an entire festival combining electroacoustics and acoustic ecology was co-organized in 1992 by Chantal Dumas and Claude Schryer.[85] This festival for "electro-acoustic ecology" took over entire parts of the city, commissioned ambulatory sound sculptures built using loudspeakers and bicycles, and featured a new work by Murray Schafer for 50 musicians occupying the whole of Lafontaine Park. The festival also made inventive use of radio broadcasting combined with environmental settings and sources. During the regular schedule of CBC-FM, ambient sound from various locations in

Montréal was broadcast without warning, in ten- to thirty-minute segments. This interpenetration of the urban soundscape with the broadcast milieu was further developed in outdoor performances composed to accompany broadcast signals received by performers and the audience on portable radios.[86]

Glenn Gould's Technological Incarnation

The core argument of this chapter has been to establish the prevalence of what I am calling an "expanded musicality" in Canada, which was articulated in Gould's prospective ideas about the future of electronically mediated music, in the contextual emphasis, both creative and scholarly, found in acoustic ecology, and in the various novel improvisatory and intermedial uses made of electroacoustic techniques. I return now to Gould's own creative work as producer of radiophonic compositions, works that blur the boundaries both between music and speech and between documentary and drama.

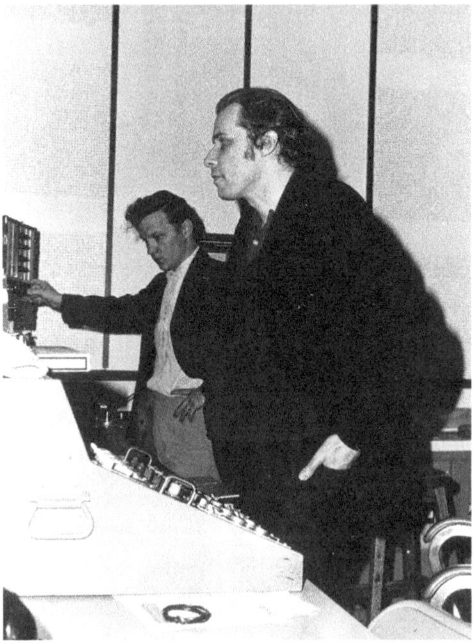

Figure 5.1
Glenn Gould (foreground) and Lorne Tulk, CBC Studio, Toronto, late 1960s. *Photo credit*: Ron Andrews. Courtesy Library and Archives Canada.

Contrapuntal radio grew out of Gould's opportunity to experiment with production techniques at the CBC after his retirement from the concert stage. Radio producer John P. L. Roberts, a close friend of Gould's, arranged for him to have an office and access to various facilities in the Toronto headquarters,[87] which he mostly frequented late at night, often working alongside Lorne Tulk, a sound engineer whom Gould had known since boyhood.[88] The CBC offered Gould a 1967 centennial commission to produce a one-hour radio program, which appeared late that year as *The Idea of North*. Gould introduced the piece, following its celebrated "trio" of overlapping voices, with these famous words:

> I've long been intrigued by that incredible tapestry of tundra and taiga which constitutes the Arctic and sub-Arctic of our country. I've read about it, written about it, and even pulled up my parka once and gone there. Yet like all but a very few Canadians I've had no real experience of the North. I've remained, of necessity, an outsider.[89]

Any current attempt to revisit this work, which many have considered a landmark in radiophonic art, must find its way through a thicket of secondary literature. Much of this writing adopts a formal standpoint, focusing on the aesthetic aspects of Gould's technique of layering speaking voices over one another simultaneously.[90] When Gould himself reflected on the challenges to listeners of hearing multiple spoken texts at once, he did so in two ways. First, he heard it as a form of collage, which disrupted the linearity of traditional voice-over documentary narrative; he was convinced that the average person can, with practice and multiple exposures, "take in and respond to far more information than we allot him."[91] (CBC radio personnel were hardly convinced; according to Roberts, Gould's work was internally condemned for being "sub-standard" with respect to technical international broadcast standards.)[92] Second, Gould explained the compositional craft he employed as a meticulously constructed musical texture, offering the example of the aforementioned trio of voices opening *North* as being shaped by what he called "continuity-in-crossover" between similar but not identically expressed ideas.[93] This motivic crossover, Gould explained, was "Webern-like," referring to the way Webern worked with small melodic cells to create a formal sense of unity lacking the harmonic moorings of tonal harmony.[94]

In addition to the formal musicological analysis which Gould put forward himself, important commentaries on *North* have been advanced which consider contrapuntal radio as a metaphor for understanding Canadian identity itself. According to University of Guelph scholar Ajay Heble, in *North* and in the two later CBC commissions forming the *Solitude Trilogy*, Gould invoked a sense of "simultaneous awareness" which can "unsettle" singular notions of identification (to group, ethnicity, nation) and "encourage us to reconceive the relationship between the whole and its parts."[95]

> Indeed, if identity, as Gould's contrapuntal method invites us to see it, is multiple, dialogic, and ever-evolving, then what is at issue is, in large part, an attack on forms and structures of authority, on constructions and representations which authoritatively claim to be able to have access to some pure, definitive, or whole truth about, say, the identity of Canada. . . . Refusing to privilege one voice or meaning at the expense of others, Gould, in effect, invites us to recognize that the combination of simultaneous voices (each significant in itself) results in a complex, but nevertheless, coherent, structure.[96]

Jody Berland, David Howes, Edward Said, and Mark Kingwell[97] are among the scholars who have written about contrapuntal radio as evocative of a kind of elevated "polyphonic consciousness," an interpretation that appears to chime well with Gould's own ideas about the communicational potential of music and media. The year after *North*, Gould prepared a one-hour broadcast on the Moog synthesizer, considering it broadly for the way it could clarify and expand on his own arguments for the role of music in the electronic age. *Glenn Gould on the Moog Synthesizer* (1968) consists of a set of alternating interviews with Wendy (then Walter) Carlos, the performer of a best-selling record called *Switched-on Bach*, and Gould's friend Jean Le Moyne, a Québecois essayist and theologian who had also worked alongside Pierre Trudeau in the Cité Libre group, which had endeavored to modernize Québec politics and, in Le Moyne's case, religion after 1950.[98]

Glenn Gould on the Moog Synthesizer (1968) differs from the full-blown contrapuntal radio works of the *Trilogy* in that Gould chose to include the on-air interviewers' questions directly in the program; he provided the interviewers with the questions in advance. At about the midpoint of the program, he overlapped the voices of Le Moyne and Carlos contrapuntally, and the ideas he chose to combine during this short passage can be heard to have a singular importance in understanding Gould's ideas about the relationship between media technology and art.

Le Moyne had a fascination with the poetics of machines from boyhood, which he spent in Montréal's Saint-Henri district enjoying "reveries" arising out of the beauty and power of steam engines, freight trains, and shipping equipment.[99] After describing his childhood reveries, Le Moyne goes on in the interview to distinguish between these nineteenth-century technologies and electrical networks, to which, in response to Gould's queries, Le Moyne attributed a new kind of "progressive consciousness," a rationality even: the man of electricity, he said, is "more intelligent" than the man of steam, also enjoying an "interior happiness." Carlos, meanwhile, explained the way the synthesizer he played on the Bach album, customized in close collaboration with its inventor, Robert Moog, allowed for fine expressive control. Carlos also told Gould that the machine had the potential to create absolutely new sounds that *"never existed in nature."* This phrase appeared just after Le Moyne introduced his own idea that the "new machines," based in electricity and networks, had the potential to supplant the old humanistic dualism, dating back to the ancient Greeks, in which speculation and direct physical experimentation were kept separate. Le Moyne stated that in the case of "superior machines," the contact with the body has the potential to in fact be "more human . . . more rational." "At a certain stage of evolution, to be in contact with an *improvisation close to nature*" becomes a more human operation, Le Moyne stated, not an alienating one. Gould pushed Le Moyne to elaborate on this by addressing the role of machines more directly in art: "Do you find it strange to connect machines with art? To have a machine between you and the work of art?" Le Moyne replied:

> Not at all. Reflecting on the musical organ, a musical machine [which has] motors for air, complicated mechanisms, and a kind of anonymity, concerning the touch. Your temperament, your style will come out in the context, contrast, registration. It means that between yourself and music, there is a distance, that of an instrument which is a real machine. I can easily conceive of bigger, more complex machines, with a greater distance between myself and the work of art.[100]

It is at this moment that Gould overlaid the two voices contrapuntally:

30:30	Carlos	But, since the resilience of the keyboard is physical, and . . .
30:35	Carlos	sort of laboring to the fingers, it makes rapid passages and certain
	Le Moyne	. . . So, I'm not at all scandalized by the idea of. . . . . the. . . . .
30:39	Carlos	. . . . decorations very difficult to play
	Le Moyne	growing presence of machine . . .
30:41	Carlos	This is the main reason why . . . *(fades out)*
	Le Moyne	between the work of art and the artist. . . . . . That's normal

When Carlos doesn't complete his sentence, after bringing up the current limitations of Moog's keyboard for certain technical passages, it is not clear whether the inaudibility is deliberate on Gould's part, or just an error.[101] But the effect, only obtainable on repeated listening, is clear enough, to put two related ideas into a visceral affective connection: the first one being about the keyboard player's challenges of articulation in synthesizer performance, and the second being about the theologian's refusal to be upset by the increasing presence of the machine between humanity and the work of art.

Later in the interview, Le Moyne went much further in expounding his philosophy of art mediated by technology. Gould opened the segment by suggesting that the "symbiosis between man and machine can only get closer." Le Moyne then stated that there is *already* a network of machines and techniques encompassing the earth.

> It is the ensemble of all the networks, the radio, television, oil, hydraulic, railway, telephone . . . so that today it is almost impossible to imagine the machine isolated from the rest. . . . There is only one machine encompassing the earth. And that has a meaning: That machine, stemming from the activity of man, is between man and nature like a second nature, offering to us its mediation. We cannot go to nature now without going through the network, just as in the spiritual [realm], we have to go through the Lamb.

Except for the final sentence, Le Moyne was voicing to Gould a near verbatim statement made by Gilbert Simondon in an interview that Le Moyne himself had made with the French philosopher earlier that year.[102] Le Moyne asked Simondon to elaborate on the concept of "networks" in the latter's treatise *On the Mode of Existence of Technical Objects*.[103] Simondon explained that networks of technical systems mediate between man and society on one hand and nature on the other. He went on to offer a list of the "ensemble" of constituent networks of the time, as Le Moyne had to Gould: roads, pipelines, telephone, telegraph, radio, satellites, transportation, and new ones to come, involving information itself, a global network of networks.[104] A few years later, Gould would himself present the same concept, of a network of networks mediating between man and nature, in an interview he made in 1971 with Artur Rubinstein. The Polish-born pianist represented an absolute contrast to Gould, as a concertizing virtuoso still skeptical about the intervention of technology in music through recording. Recording, for Rubinstein, still served just to document a live performance, to the degree that even hearing spoken voices reproduced through records or the radio still frightened him. Gould then introduced Rubinstein to Le Moyne as a poet and theorist of technology, beginning with a paraphrase from Le Moyne's interview with him, to the effect that "there should be no contradiction between technology and the humanities, in particular, theology." Then Gould tells Rubinstein,

> Technology has now given us something like a network—a radio network, a television network, an oil network, a railway network, a communications network of all kinds—and this network has circumscribed the earth in such a way that we can no longer go into nature, *we can only go to nature through the network*. But when we do that, we realize that technology exercises a great charity on our lives.[105]

As Gould understood the idea, charitable technology initially offered a way to provide many more people with musical experiences than could be reached in live concerts. But charitable technologies could also offer a means of moral improvement: people might be able to attenuate their competitive hyperindividualism and base instincts if distanced from one another through the sort of networks Le Moyne and Simondon articulated and Gould endorsed to Rubinstein.

Le Moyne's theology of benign technology goes even further than this, right back to scriptural sources. In the 1968 CBC program on the Moog synthesizer, Le Moyne concluded his statement that going "through networks" was like going through the Lamb in the spiritual realm. Speculating on the spiritual meaning of the "process of quantification," Le Moyne then says:

> There are a lot of things we have to quantify that we don't dare to yet. Probably almost everything of man can be quantified in some way; that is *Incarnation*. . . . Christ [did] not hold himself jealous of the rank that equaled him to God, [when] he went down to became man.[106]

Le Moyne admits that connecting the thought of Jesus's Incarnation with the wider implications of quantification can appear somewhat farfetched, but it is clear enough from the context what he is getting at. The insertion of man into the networks of technology is comparable to Christ's obedience in coming to earth and becoming man; this "materialization" of spirit (or number) into flesh appears to have been a deep-seated intuition in this oral exposition of Le Moyne's humanist theology of machines. "That obeisance is exactly the insertion of man in the determinism of his materiality. We have not yet explored the extent of that materiality, it is much more than we think. There is a kind of Christification going on in the machinic world, in technology."[107] This is the final word Gould gave to Le Moyne in the program: Gould then had engineer Lorne Tulk fade in a synthesized performance by Carlos of Bach's "Jesu, Joy of Man Desiring."

For the second subject of the *Solitude Trilogy*, Gould traveled to Newfoundland to explore the effects of isolation and modernization in that province's outport fishing villages. For the third in this series, *The Quiet in the Land*, he returned to the theme of spirituality, interviewing members of a Mennonite community in Manitoba. By then, Gould was working with the newly available multitrack tape recorders, permitting a much more precise way of handling the "Webern-like continuity-in-crossover" that he had earlier sought to effect with less accurate tools in *North*. Two of the principal characters in *The Quiet in the Land* are a pious minister, Father Neufeld, and a skeptical academic, Clarence Hiebert. In a significant contrapuntal passage halfway through the program, Father Neufeld is discussing Passover and the Last Supper. Gould then overlays Hiebert's voice in a precise counterpoint:

32:25	Neufeld	Jesus lived in a time of conflict. . . . . and it seems. . . . .
	Hiebert	. . . . It appears to me that the. . . . .
32:29	Neufeld	to me that there was an inevitableness that in that context. . . . .
	Hiebert	idea of the Incarnation of Jesus. . . . . his coming to the earth
32:35	Neufeld	. . . . one such as Jesus taking the position that he did,
	Hiebert	and being among people. . . . . is the newer and better understanding
32:39	Neufeld	. . . . was bound to end in death.
	Hiebert	of what the "in the world but not of the world" concept is.

Being "in the world but not of it," a kind of Mennonite credo, was a phrase often used by Gould too. In this contrapuntal passage, Gould connected it with the Incarnation, in a complex resonance with Le Moyne's theology, while associating the phrase with the inevitability of Jesus's death as well. University of Toronto philosopher Mark Kingwell has aptly called Gould a "progressive and anti-progressive at once": a fierce opponent of the zeitgeist while also being himself one of its "most interesting expressions."[108]

In his final contrapuntal radio composition, the little-heard *Glenn Gould: A Radio Fantasy*, released in 1980 on Columbia records, he returns to this theme, in a passage that can be heard now as Gould's final word on contrapuntal radio and a summation of his own personal credo. An inveterate mimic and impersonator, Gould played many roles in the *Fantasy*, indeed becoming to an absurd degree what philosopher Geoffrey Payzant suggested he sought to be his whole life: a "Poly-person."[109] He voices, through the role of Dr. Klopweiser, a German musicologist, this self-assessment of *The Solitude Trilogy*.

Klopweiser: There is always the theme of isolation, the characters are removed from society in some way. The drama is created by their concerns, their uncertainties, their doubts about what alternative states might be available to them. . . . There is a very strong statement against the zeitgeist, sometimes it is quite obvious, sometimes it is subliminal. I have the feeling these characters are not quite of their time. They do not wish to be pulled, to be tugged by the gravity of the zeitgeist.
Gould: They want to be in the world but not of the world. . . . Because, I don't think one can benefit from isolation in whatever form, whether it's in an Arctic outpost, or a Newfoundland village with no road to the outside world, or a religious community, which are the main metaphors we used in the *Solitude Trilogy*, I don't think one can benefit from that without coming to terms with the zeitgeist, without deciding that its tremendously tyrannical force has to be overthrown in one's life before one can, you know, learn from such an experience.[110]

Gould's ambivalence, presented here through the sharply contrasting tenor of his enthusiastic visions for technology in music alongside his anticonformist aversion to the nostrums of inevitable progress, has been diagnosed by some critics in clinical terms; indeed, a whole literature exists about his mental health and physical failings.[111] My purpose here has been to examine it more productively, as a necessary resource that fed his desire to compose in a new way, not using the traditional materials of music, like notes and harmonies, but using speech transformed and structured in musical ways. Further, it rings together with a sense of prescience, as the idea of ambivalence as a cultural resource, so central to Gould's creative stance, has been highlighted in recent years in writings about the challenges to innovation in a world facing a "fragile future."[112]

Coda

Within a few years of Gould's death in 1982, key elements of the arguments he made twenty years earlier for music in an electronic age came roaring onto the scene, some even on an industrial scale. The synthesizer that had so fascinated him in the hands of Wendy Carlos—an instrument that had been customized by Robert Moog to draw out the specific contours of Bach's music—was soon being mass-produced in digital formats designed to be interconnected using a network protocol for musical information transmission. The Musical Information Digital Interface (MIDI) joined personal computers with synthesizers, helping to spark a burst of sales in hardware components as well as new markets for the digital sounds themselves and a new class of software applications for composing, arranging, mixing, and mastering music.[113] The widespread adoption of these music-making devices was controversial in some quarters; in academic and experimental music communities MIDI stunted the imaginative scope of electronic music by reducing it to numbered notes with hard-edged on-off triggers, even if the sonic properties of these triggered notes could be modulated in sophisticated ways.[114] For Canadian communication scholar Paul Théberge, this standardization represented the "incursion of capitalist relations upon [musicians'] creative practices at the most fundamental level," forcing them "to adopt increasingly mediated forms of communication with one another."[115]

While McLuhan forecast an infinite jukebox of the world's music, past and present, the actual realization of this vision came fifty years later mediated through a system of data analytics that delivers "recommended" musical commodities just as fixed and frozen in their formats as those Gould had decried as inducing mass conformity by turning the live act of performance into the mimicry of replay. In a similar vein, Gould's idea of a charitable network of humane technologies demands to be revisited through the sober eyes of a particularly enveloping form of platform capitalism that now encompasses music consumption.

Canada's expanded musicality during the 1960s–1980s was complex and wide-ranging. It was formative internationally, notably in the Canadian contributions to the birth of acoustic ecology and soundscape studies. The ongoing richness and impact of its improvisatory music-making (often electroacoustically inflected) and its scholarship on improvisation are also exceptional.[116] These innovative contributions in the musical idiom, from the discursive to the aesthetic to the technological, exemplify Canada's alternative technological ethos as both creative method and model of interdisciplinarity.

6 Two Faces of Cultural Networking

The year is 1969, and in Ottawa, the communication-besotted capital of Canada, understanding new media was the order of the day. The country's recently elected, coolly modish leader, Pierre Trudeau, corresponded regularly with Marshall McLuhan on the powers and dangers of electronic communications.[1] New content regulations were put in place by the broadcasting regulator, the Canadian Radio and Television Commission, to encourage more production of domestic cultural programming. Trudeau set in motion a task force to study telecommunications from a broad perspective, not limited to technology and regulation, but including its cultural potential and social impacts. The resulting report of the Telecommission, *Instant World*, preceded by seven years the more well-known French effort, the Nora-Minc Report promoting the "computerization of society." The Canadian report notably emphasized a new kind of human right that would accompany the new digital media on the horizon, a "right to communicate" as fundamental as the freedom of assembly or speech.[2]

In his April 11, 1969 letter to the prime minister, McLuhan, then at the height of his world fame, railed against the very idea of thinking of culture as units of programming to be transmitted from sender to receiver over modern broadcasting channels. Television networks, including the Canadian Broadcasting Corporation, would collapse, he warned, unless they started to think of their main business as the coming "knowledge industries"; instead, they were stuck in thinking about "packaged programming, nineteenth-century style."[3] Elaborating further on the point in a 1971 letter, McLuhan recommended that Trudeau read his latest, though apparently never published essay, "Putting On the Media: The User Is the Content." In it he explains:

> The user is the content of any medium or environment. . . . We are the content of anything we use, if only because these things are extensions of ourselves. The meaning of the pencil, or the chair I use is the interplay between me and these things. Again, the message of these things is the sum of the changes that result from their social use. Thus I have added two things to "the medium is the message," namely the content and the meaning.[4]

By clarifying the commonest of catchphrases associated with his 1964 book *Understanding Media*, McLuhan was drawing the prime minister back to the essential point of his first explicit article on this topic, which appeared in 1954 as part of the second volume of *Explorations*, the journal McLuhan co-edited with Edward Carpenter.[5] Media, McLuhan proclaimed, need to be understood as "art forms," rather than in terms of the then-prevalent understanding of communication as a mere matter of the transmission of information. The "naïve content-view of culture prevents us from directing serious critical attention to the media, old and new, as art forms."[6] Specific media have distinct life cycles, pressing as they become familiar and widely used toward a "maximal realization"; he provides the example of painting which, after the invention of photography, was relieved of "having to be concerned with narrative and anecdote," thus becoming a "vehicle of sheer expression."[7] In his 1971 letter, McLuhan suggested that the Canadian Radio and Television Commission's director, Pierre Juneau, ought to read the suggested essay to gain an understanding about the real problem of "Canadian content," which would be "drastically simplified" by understanding its meaning not as something transmissible as such, but rather as a matter of "interplay or dialog between Canadians." By insisting on the user as the content of modern telecommunications, in terms consistent with today's social networking platforms, McLuhan was implying that public policy on culture and communication needed an entire rethink: "It is unfortunate that the CRTC ever involved itself in the question of content, especially since it does not understand the nature of media at all, except as hardware."[8]

Also in 1969, at the National Research Centre in Ottawa, the first Canadian seminar on Man-Computer Communications was sponsored, with keynote speaker J. R. C. Licklider from the Advanced Research Projects Agency of the US Department of Defense. Licklider's presence in Ottawa is noteworthy for the ripples it caused within the research community; though the Ottawa speech was not published, the speech was likely very similar to the one Licklider gave in April 1969 to the Spring Joint Computer

Conference.[9] It must be recalled that while he is historically best known for having sponsored the fundamental research leading to the internet and interactive computing, Licklider was an acute perceptual psychologist, and had as a younger man been an active participant in the interdisciplinary Macy Conferences, alongside such figures as Norbert Wiener, Claude Shannon, Margaret Mead, and Gregory Bateson.[10] Given this intellectual history, then, it is not surprising that Licklider's primary concern at the outset of the Ottawa address was to differentiate between the potential role of computing as communication and as representation.

> As the history of art shows, the representation with the greatest verisimilitude is not often the best means or medium of communication. Good graphical communication is more a matter of linguistic expression than it is of pictorial reproduction. Unfortunately, we in the computer world do not know much about the language [of representation], and unfortunately not many are trying hard to find out about it. Obviously, it does not have much to do with programming languages. It probably has something to do with data structures. But mainly it is the art of expressing ideas through configurations and manipulations of signs and symbols.[11]

Licklider's thinking from 1969 resonates strongly with Jonathan Sterne's recent rethinking of media historiography. The established historical narrative of media is one of increasing verisimilitude, a forward march toward greater definition in terms of display resolution and immersion. While the "claims about increased definition or the pleasures of immersion are in many cases true," Sterne points out that "they are not the only story."[12] A second communication history has yet to be written, he argues, one concerned not with resolution or fidelity, but with how sensory experiences can be *compressed* in ways that, as Licklider was suggesting in 1969, can produce more effective or relevant representations for a particular purpose. In Sterne's general formulation, "compression is the process that renders a mode of representation adequate to its infrastructures. But compression also renders the infrastructure adequate to representation."[13]

It is within Sterne's alternative history of media as compression that the local impact of Licklider's visit to Ottawa in 1969 can be best understood. Immediate reactions to Licklider's address focused on his suggestion that under then-current economic and technical limitations of computing, attention should be given to the development of new kinds of "ideographic writing," that is, not communication that aimed at verisimilitude, but rather effective new forms of representation using computational "manipulation of signs and symbols."[14] Whether inspired by Licklider's speech or not, the reverberations of this thinking can be detected four years later at the same Man-Machine Communication seminar series in a lecture by Gordon B. Thompson, a communication engineer and futurist author then employed at Bell Northern Research in Ottawa. Thompson's concern in 1973 was to advocate for Canadian leadership in the development of editing and assembly systems for ideographic communication, which would "greatly speed and ease access to stored human experience."[15]

Read against Sterne's recent analysis of compression, Thompson's design proposals reveal the clarity with which Licklider's vision of computer-based graphical representations contrasted with McLuhan's conception of media as "art forms" in which the "user is the content." For McLuhan, content itself is not a transmissible function, but rather an interplay or relation conditioned by the properties of the various media themselves, in this case, the properties of real-time interconnectivity. By contrast, for Licklider and in the subsequent "compression-based" thinking that followed in the wake of his Ottawa visit, the relationship that mattered was how well new creativity can be supported within the very constraints of a format or communication infrastructure. This chapter will follow this basic distinction in presenting two faces of cultural networking in Canada, between about 1967 and 1986. The first of these aspects of cultural networking adopts what has since been called an "aesthetics of telecommunication," in which the "role of artist and audience become intertwined . . . and the very notion of meaning . . . is revised by the experiential setting of communications."[16] The second face of cultural networking centers upon a sophisticated, technically brilliant attempt to deploy a national videotext system—Telidon—based essentially in its design on an advanced protocol for content compression.

The first story will be organized around a group of artist-initiated networking initiatives, which mainly grew out of the efforts of experimental artists working interchangeably with "old" and "new" media. This approach was informed by conceptual art, systems and cybernetic theory, and in some measure by McLuhan's ubiquitous writings on media. It was supported by public grants for employment assistance and later by a full-fledged program of the Canada Council for the Arts. The artistic efforts that gave shape to this approach spanned network-based thinking using established means of correspondence, like postal mail, telephones, and low-powered radio broadcasting, as well as newer ones like digital networks, facsimile transmission, and video conferencing. The combination of old and new media in this bottom-up artist-initiated approach to networking contrasts sharply with the Telidon story, which reveals a government-led, techno-economically driven approach, spearheaded by the federal Department of Communications out of one of its flagship national research labs. Despite lavish funding, both public and private, and widespread praise for its technical advances over the French Minitel and other already-deployed digital teletext and videotext networks, Canada's prestigious Telidon effort failed to ever find a significant public uptake and was discontinued by 1985.

These two contrasting approaches structure the story I want to tell in this chapter. In terms consistent with the ways artists and engineers worked creatively with early computing in previous chapters, the artist-led networking initiatives were imbued with the alternative technological ethos of making creative, often subversive use of systems and tools already available, if sometimes in still-nascent forms. Counterposed to this was the more futuristic "build it and they will come" orientation of the government planners of the Telidon project, whose approach demanded a new, unfamiliar kind of linguistic competence without really thinking through the implications of this fact for users and widespread adoption. This competence required a different set of skills, oriented around symbol manipulation and involving a powerful coding scheme adapted to the economic limitations of low-bandwidth data lines.

When artists were exchanging live images over telephone lines, each frame taking some eight seconds to appear, the aim was to explore the affective experience of electronic visual connectivity. For these artists, the "lack" in the available technologies was not an impediment; they simply worked with the tools and systems they had at hand. Artists did approach myriad systems of communication as modes of globally connected experience, sometimes anticipating what they could sense as coming forms of cultural and social life. Hank Bull, one of the Canadian pioneers of telecommunications art, which flourished from around 1979 to 1986, would later call this an "experience of mediated engagement," anticipating something that would become commonplace decades later with the arrival of the consumer internet and ubiquitous mobile devices. Making reference to some of the intellectual tendencies of postmodernism, Bull notes that these explorations of "the disembodied, the virtual, and the ephemeral made sense in the era of the simulacrum, the collapse of meaning, and 'death of the author.'"[17] They were also continuous both with the anti-object thinking of conceptual art (which saw a pronounced development in Canada during this period),[18] the cybernetic network-based thinking of system theory,[19] and the media discourses of Canadian communication thinkers. As Bull recalls of this heady time, when performance and conceptual artists, musicians and poets were connecting in real time with each other around the world, artists then were playing their common role in all eras: "to feel the emergence of new systems of social organization, to predict and embody the coming forms of human consciousness."[20]

Telidon can be seen today both as a financial fiasco, a near textbook case of how *not* to inject public stimulus to kickstart new high-technology services, and as an instructive indication of the two separate technological art cultures that coinhabited the Canadian scene at the time. The distance between the two approaches to telecommunications networking profiled in this chapter was gaping, in spite of past successful bridges that had been built between cultural and scientific institutions—the NFB and NRC, for example, as seen in chapter 3—and the progressive communication thinking that sprang from the deep pockets of Canadian policymakers during the 1960s and 1970s. Telidon's sophisticated data protocol, designed in the late 1970s, anticipated some aspects of what would become the browser applications for the World Wide Web in 1992. It was initially conceived to enable the creation of a "common communication space" for sharing compressed images, texts, and sound over then-prevailing low-bandwidth networks. But the conception of *what* would be communicated over Telidon by its government backers and corporate information providers was fatally impoverished, so that, by the time it was made available to users, it consisted essentially of transactional information services accessed through alphanumeric keyboard commands (not the emerging graphical user interfaces of the newest microcomputers). In addition, unlike the French strategy of subsidizing the production costs of cheap Minitel terminals, Telidon required expensive hardware to be purchased by end users, a cost that was never well justified to consumers while lacking much in the way of attractive information services. Some insightful explorations did occur around Telidon's compressive infrastructure, and their incomplete realizations are recounted here in the spirit of a media-archaeological recollection of incomplete, stillborn, or imaginary media experimentation.[21] Nevertheless, when the actual formal artist trials of Telidon systems finally began in 1983, some six years after the system was designed, the whole project had come to appear belated and pedestrian, certainly as compared with the experimental telecommunication projects initiated by artists as presented in the next section, or even with the more radical creative uses of the simpler, first-generation text-based systems, like Minitel, in other countries.

Artist-Initiated Networking in Canada

Artist-run centers emerged intensively as an international movement during the late 1960s and 1970s, signaling "a move away from art-making's preoccupations with the renovation of existing models of institution to the construction of new ones."[22] These emergent artist-run institutions often enabled engagement with new media forms (like video art, electronic music and photography), genres (like performance and installation art), and social movements (like feminism and minority rights in the arts) that had little support at the time from mainstream museums and the commercial art market. The Canadian cast of this international tendency was from the start different from other countries, owing to the vigorous and sustained role played by state funding in support of artist-run centers as a crucial component of *national* cultural development. Curator and art historian Diana Nemiroff linked geography, cultural policy, and political economy together when she argued in 1985, some fifteen years into their history:

> The artist-run centres represent a distinctively Canadian response to the problems of exposure and distribution of contemporary art in a large, sparsely populated country that lacks a viable market economy of art even in the main metropolitan centres. Thus, the network of artist-run centres can be thought of as a different economy of art, not necessarily antagonistic to a market economy, but organized around other principles than the laws of supply and demand.[23]

For the first decade of their existence, these collectives and "parallel" spaces constituted their own public, as A. A. Bronson notes, which rose out of the relative thinness of Canadian national identity, the permeation of Canadian airwaves by American popular media, the impossibility of competing with New York as world art capital, and the "McLuhanistic policy making of the Canadian government in the late 1960s and 1970s."[24] Instrumental among these federal policies were job creation programs that were structured not as public works or market-oriented retraining, but rather as community and cultural development and responding to individual initiatives. Introduced in 1971 with the substantial budget of $180 million, the Trudeau Liberal government's Opportunities for Youth and Local Initiatives Program seeded artist collectives as well as new cultural institutions across the board, with an aim to not only create jobs but also "public goods—and to promote public good—in a democratic system of civic as well as economic production."[25] By the late 1970s, Canadian artist-run centers constituted a network, supported by Canada Council funding, whose spokespersons self-consciously understood its role as being nothing less than "the proven primary providers for the future of Canadian art and culture. . . . The artist-run centre is as important to the health of the culture as the preservation of heritage and tradition."[26] By the later 1980s, the Canadian artist-run network focused attention on the "democratization" of services, a shift from "oppositional to identity politics"[27] concerned with redressing racism, homophobia, and sexism, during the same years that in the United States the political right was vociferously targeting radical art, artist collectives, and progressive institutions during the "culture wars."[28]

Two other expressions of artist-initiated networks, one philosophical and the other technological, also animated this burgeoning Canadian federation of artist collectives. The first, philosophical expression emerged out of the proposals of Robert Filiou, a French artist associated with the Fluxus movement, Vancouver's Western Front, as well as various Québec centers and publications. Radically decentralist in spirit, Filiou also opposed the very concept of the avant-garde, with its emphasis on a dialectic of progress and obsolescence, and proposed instead the concept of the Eternal Network.[29] In Filiou's nongeographic network of the mind and spirit, each artist is individually a node, and metropolitan centers are replaced by "places": "Nobody can tell us where the place is. Where we are is where things are taking place and although we may need to meet at times or gather information at certain places—the network works automatically."[30] The Fluxus conception of art as a way of life is a matter of ongoing mutuality and conversation, a process which translated smoothly into actual telecommunication-based performances and experiments.[31]

A second, technological expression of network thinking is, appropriately perhaps, more difficult to connect to a single figure. By the late 1970s, some members of the artist-run network put forward a proposal to create a technologically defined system that would become the basis for a "creative approach" to new communication media, reflecting the idea that for many of its members, the "medium *is* their artform."[32] The network would both serve as an "Agency," an organizational and touring utility, as well as providing for new kinds of sociality understood in creative, not just utilitarian, terms. This proposal was met with some skepticism, particularly when it was presented somewhat ironically as a quasi-bureaucratic "Living Museum" functioning as an "Apparatus."[33] Victoria-based Bill Bartlett, a primary instigator of artist-initiated telecommunication in Canada, argued for the need to appropriate industrial technologies for humanistic aims, to demystify and explain their potential at large. Realistically accepting that "mechanization, convenience and accessibility are the passwords for the future," Bartlett suggested that artists' work will undoubtedly mirror these preoccupations: "Artists' involvement in shaping technological development, in assuring that this development reflects humanistic concerns, may well alter the directions we seem to be heading in."[34] Such rhetoric proved to be controversial, inciting a response from some of the network participants who chose rather to perceive their role as "seedbeds" and "providers" of new art materials for the mainstream art market. Diana Nemiroff, for instance, sharply noted in 1983 that "the impractically grandiose conception of a computer-based data network represents the recurrent desire to appropriate the outward symbols of power of the profit-making communications empire. . . . It is a gloriously perverse attempt to institutionalize the anarchistic, decentralized 'eternal network' by installing a head office."[35]

Sat-Tel-Comp, *Wiencouver, The World in 24 Hours*, the *Planetary Network*

Bill Bartlett was a recent art school graduate from Los Angeles when he moved to Victoria, British Columbia in 1970, at the height of the Vietnam War.[36] As a student, he had participated in West Coast projects organized by Experiments in Art and Technology, the international group that continued to support collaboration between artists and engineers following its iconic 1966 launch event, *9 Evenings: Theatre and Engineering* in New York City. With his experience working on the Pepsi-sponsored pavilion for the 1970 Osaka world's fair,[37] Bartlett turned to media technology as a tool for community development during the mid-1970s, in his capacity as director of one of the charter Canadian artist-run centers, Open Space. Calling

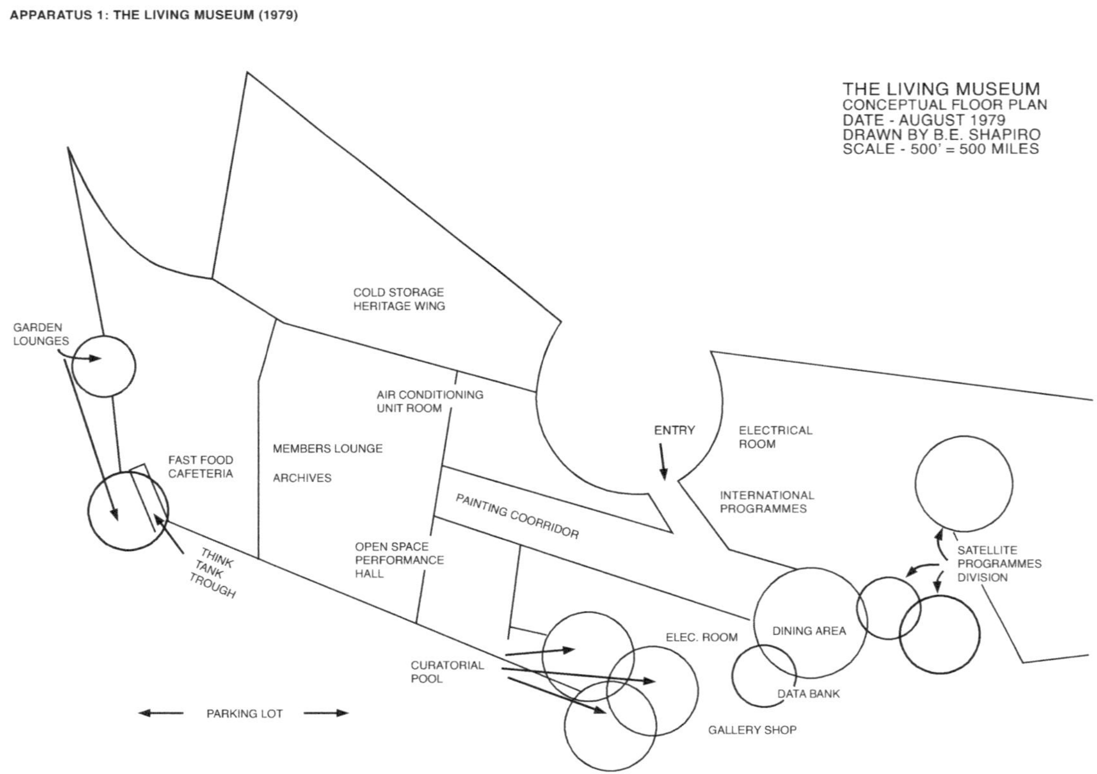

Figure 6.1

Apparatus 1: The Living Museum conceptual floor plan. Adapted from Barbara Shapiro, in Lewis, "The Living Museums Data Network," *Parallelogram* 3 (1979). ANNPAC/RACA, Toronto, 1979. Restructuring: reversing the hub-and-spoke model of a national museum within a revised Canadian political geography. Imagining a federation without a center. Disguise: maintains an idea of an artistic network where art administration and artist-run centers are also absent.

his curatorial concept "Collaboratory," fusing collaboration and laboratory, Bartlett brought together artists, engineers, corporations, and government agencies around uses of video, satellite, fax, computers, and print, "in an attempt to create ideas and concepts, as well as make art."[38] The signature technical method was dubbed Sat-Tel-Comp, standing for satellite-telephone-computer, which in effect came simply to mean slow scan TV, or SSTV. SSTV sampled a video image into computer memory every few seconds and sent it through telephone lines as an audio signal; reception took from 8 to 60 seconds, depending on desired picture quality (from 128 lines black-and-white to 512 lines color). "The result was a time-lapse series produced at intervals, like a very slow animation. The digital scan moved from the top to the bottom of the screen, line by line, and when the total image was scanned out, it was held until replaced by the next scan."[39] As understood by Bartlett at the time, drawing his vocabulary from German artist Joseph Beuys, the conjoining of communities through telecom was a kind of social sculpture, aiming "to tie together centers from around the globe into one 'piece.' . . . Our intention is real-time broadcast of interdisciplinary activities and process."[40] By 1980, Bartlett was a key instigator of connectivity experiments, working first at Open Space and then at a separate media production organization founded expressly for this purpose, recruiting and training curators, historians, artists, and critics from around the world.[41] Bartlett soon passed the mantle on to other artists, finding the costs and organizational challenges too onerous, and turned to a simpler form of information transmission as postmaster in the British Columbia village of Saanichton.[42]

The Western Front Society in Vancouver, another of the earliest artist-run centers (founded in 1973), became one of the global hubs for hybrid Sat-Tel-Comp-style networking. Growing out of a free-spirited West Coast energy, according to artist-curator A. A. Bronson, Vancouver artist centers had initially combined an interest in new dance, performance (centered at the proto-ARC Intermedia in the late 1960s),[43] as well as enthusiasm about communication and new media.[44] Unusual even among other interdisciplinary centers, the Western Front programming spanned visual arts, experimental music, performance, and was early to introduce video production facilities.[45] Hank Bull at the Western Front picked up on and developed Bill Bartlett's initiatives, notably in a series of happenings and actions using SSTV, computer conferencing, and telephone audio, joining Vancouver and Vienna, Austria, where his ongoing collaborator was Robert Adrian, a relocated Canadian with close connections to the Austrian state-run Kunstradio. According to Bull:

> *Wiencouver* is an imaginary city hanging invisibly in the space between its two poles: Vienna and Vancouver. Seen from Europe, both cities are at the end of the road, one on the Pacific rim of North America, the other just 65 km from the soviet bloc. They are each on the edge of the art world's magnetic field, able to observe from a distance, and equally able to turn the other way, one toward the far East and one toward the near East. Vienna and Vancouver are wealthy, regional cities with international perspectives. This, coupled with their linguistic and historical differences, makes them ideal correspondents.[46]

The *Wiencouver* two-way exchanges amounted to an "experiment in interactive television," according to Adrian,[47] combining live sound over the phone and SSTV, and helped prepare the expertise and networking style for more expansive experiments in global connectivity over the next few years. One of these, spearheaded by Adrian himself, was *The World in 24 Hours*, part of the Ars Electronica festival in 1982. This global network event used SSTV, fax, computer mailboxes provided through ARTEX,[48] and telephone sound, successively connecting fourteen cities on four continents.[49] For Adrian, the eclectic mix of readily available and leading-edge systems, combined and recontextualized, was decidedly a low-tech approach, and the aim was inclusivity: "Telecommunications projects which do not seek a global participation are little more than parlor games."[50] The 1986 Venice Biennale sponsored an even more englobing project, involving similar means: the *Planetary Network*. One of the co-curators of the Venice event was Roy Ascott, who after his short tenure as art college administrator in Toronto became an enthusiastic proponent of telematic experimentation as a mind-expanding cultural horizon. The cybernetic vision he had tried and failed to implant fully at the Ontario College of Art in Toronto was, by 1984, ready to be realized in what he preferred to call telematics (a word adopted into English from the French *télématique*), and grounded in his understanding of the essential link between the computer and networked communication.[51] From Toronto, OCA instructor Norman White was a contributor to the *Planetary Network*, and along with Ascott, Tom Sherman served as one of its co-curators.

In Adrian's retrospective account, the *Planetary Network* was the culmination and conclusion of this whole period: "The networking of Personal Computers in BBS's and the increasing presence of Fax and other telephone peripherals in offices made ARTEX and [similar telecom] projects superfluous. The pioneer days were over."[52] Hank Bull, one of the connective threads running through this period in Canada and internationally, commented that the mere fact of connectivity was astonishing: "to hear a voice, read a message, or see a face on the screen, beamed from afar, seemed like some kind of miracle." The preparation required for such fleeting epiphanies was laborious, and the equipment unreliable and expensive, to the degree that for Bull, "the matter of content became an afterthought."[53]

Vera Frenkel's *String Games: Improvisations for Inter-city Video*

Considering the "content" of an artwork as a provisional afterthought could hardly be more antithetical to the approach toward communication technology taken by the internationally renowned Canadian multidisciplinary artist Vera Frenkel.[54] Born in the heart of central Europe in 1938, marked by the tragic birth date of November 9—the Nazis' infamous *Kristallnacht*—Frenkel came to Canada as a young girl and trained in anthropology at McGill University. This choice of discipline informed her ongoing intellectual stance, which always sought to understand the social role of art and art making in the broadest sense: "I wanted to be able to look at the world I live in with some perspective, or, more accurately, from multiple perspectives."[55] During her time in Montréal, she was shaped by the exuberance of Expo 67, the breakthroughs in brain science led by Penfield and Hebb at McGill, and by the feminist perspectives on culture taken in anthropology by Margaret Mead and Ruth Benedict. In 1973 she received an offer to exhibit her visual art, and to present her writing, at several galleries simultaneously. Frenkel then disarmingly proposed the fresh idea of simply connecting the galleries together by video hookups and screens. After inquiring with Bell Canada, she was told the cost of laying cables would be prohibitive, but also learned about a new point-to-point dedicated videoconferencing facility soon to be opened, connecting Montréal and Toronto. Bell was then just testing out the system, needed trial users, and was willing to let her visit the Montréal studio to check it out.

After seeing the fixed seating configuration of the studio, which consisted of five stationary chairs for participants and three fixed cameras, Frenkel conceptualized a scenario to serve as a challenging test of the corporate videoconference setup, repurposed as an improvisatory social encounter. *String Games*, significantly subtitled *Improvisations for Inter-city Video*, was presented in November 1974. Taking each of the five seats as a finger and joining the two cities as a virtual pair of cooperating hands, Frenkel devised *String Games* initially as a play on the cat's cradle—a children's game in which varied topological shapes are formed from a single loop of string threaded between the fingers of two hands. Frenkel provided each of the "fingers" enacted by a separate person with a set of instructions on which to base the changing configurations of the imaginary cat's cradle. These prompts consisted of an item selected freely by each participant-performer from the following ordered list: a number, a letter, a word, a name, a sentence, a fragment of a poem, a visual image, a gesture, and a sound. Frenkel's idea was that

> each player would learn the movements of a particular finger in all the figures and could contribute to the improvisation in the same order as that finger would manipulate the string, [as] if we were playing with a string loop rather than a video connection at a distance of 350 miles. The eight string figures meant eight sets of movements for ten players to learn. These sets of movements were the ordering principles for the piece.[56]

Frenkel went on to become an eminent video and installation artist and one of the earliest "net artists" in the 1990s, known primarily thorough a body of work distinguished for its concern with issues of migration, dislocation, displacement, and memory. Frenkel has acknowledged the importance of her role as the stranger, which "however complex and sometimes difficult, served as a useful antenna in my own work."[57] Two of the works for which Frenkel is most widely acclaimed are . . . *from the Transit Bar*, for Documenta 1992, a complex video installation functioning also as a working piano bar, and *Body Missing* (1994–2008), which exists in various media formats including video, installation, and the internet. Both are suffused with memories of mid-twentieth-century trauma and the contemporary experience of exile and migration; both are shaped as complex, open-ended multi-narratives.

String Games has also been read as a profound meditation on memory, presence, loss, and absence by subsequent critics examining the traces of the performance through video documentation thirty years later. For Toronto critic Dot Tuer, it is crucial to understanding *String Games* against the highly charged background of Frenkel's later work, which is "beyond the frame of new media."[58] With this phrase, Tuer expands on a previous appreciation of the same work by Canadian art historian Caroline Langill. Langill effectively makes the case for including *String Games* in the canon of so-called "new media art" by noting in it the common tropes of rules, emergence, and interactivity.[59] Tuer sees something more to uncover beyond these formal features that have been overlooked: a certain "pathos," beyond the "indeterminacy" of play, that is grounded in a "metaphoric loss of presence" and comes out of the necessary substitution of the letters, words, names, sounds, poems, etc., for actual persons and actual string. Frenkel herself offers a self-assessment at once formal and conceptual when she states that "the piece was located at a nexus between two formats; a very old universal one, the string game, was re-enacted and transformed via a newly developed one, video transmission."[60]

It is precisely at this nexus, between old and new and between the predetermined and the unprecedented, that *String Games* can be seen not as indeterminate play but precisely as an *improvisation*. It is well to recall how British analytic philosopher Gilbert Ryle defined the word: "Improvisation is, at the least, the engaging of partly trained wits in a partly fresh situation. It is the pitting of an acquired competence or skill against an unprogrammed opportunity, obstacle or hazard."[61] Adopting this definition, the instructions provided by Frenkel are less rules than a kind of quasi-notational palette, a set of "trained wits" providing the basis for exchange and group construction, in a manner ordered by the eight spatially defined virtual shapes of a classic children's game. "Indeterminacy," in the sense often associated with randomness and the teachings of John Cage, is thus less at issue in *String Games* than improvisation as such. Furthermore, the participants were marked by the letter X on their fronts, a sign for Frenkel of both erasure and presence, therefore working as an indicator that oscillates in meaning between presence and absence, and which thereby destabilizes and makes the temporal quality of immediacy in the intercity engagement far from transparent. Though akin to the "eternal present" time occupied by the telecommunication art pieces described earlier in this chapter, the pathos identified by Tuer in the improvisatory process of *String Games* urges us to consider the work in more complex temporal terms as well: a "mediated engagement" using remote connectivity that varies the aperture of the present at both of its edges, a present which is itself constituted by the multilayered blend of immediate actuality and virtual imaginary.

The artists presented in this section used a variety of means to support their engagements with telecommunication networks, but they shared a common sensibility, one that can be summed up in terms of interconnectivity as play, globalism, and immediacy in apprehending the process of communication itself.[62] The chapter thus far has surveyed artist-initiated networking in Canada in three parts, with narrowing focus, starting with the network of Canadian artist-run centers themselves, then presenting telecommunication events with an international cast growing out of these Canadian centers, and concluding with a single seminal work by an artist not otherwise associated with the art-and-technology tendencies of the time. In none of them did we find much interest in, or necessity for, the kinds of issues raised by Licklider in Ottawa in 1969, regarding the use of efficacious symbolic schemes for compressing and transmitting meaning using the new telecommunications technologies. The following section is about exactly this concern.

Telidon: Another Panacea that Failed

As noted in the introduction to this chapter, the predominant narrative of linear progressive media history is one of increasing verisimilitude. Compression—the adaptation of expressive intentions to the finite constraints of a given infrastructure—offers a compelling alternative to this historical narrative. As Jonathan Sterne points out in his studies of the MP3 format, and his subsequent sketch of a "loose history" of compression, new protocols can lead to new infrastructures too; for instance the MPEG standard enables satellite radio and the DVD, so that "it is not just compressions adjusting to infrastructures, but infrastructures modified by the phenomenon of compression."[63] Limited definition, which deliberately employs condensation of thought and language, can nonetheless produce "particularly intense modes of experience."[64]

Both the slow scan TV, fax, telephone, and email media mix of the early art-telecommunication performances and the two-way video conferencing infrastructure used for Frenkel's *String Games* appear to have produced such kinds of intense experiences for their participants. Both were also cases of adjustment by the artists of the expressive range and intent of their performative actions to the telecommunication infrastructures that they deployed—in each case, systems were obtained and used as given but not adapted technically as such. As McLuhan might have observed, the users surely were the content, particularly in projects like *Wiencouver*, but also in Frenkel's subtle understanding of how to animate the intercity televisual space using a gamelike structure for improvising a series of virtual cat's cradle shapes. However, these cases are different in principle and kind from the way an infrastructure based on the compression of symbolic coded representations operates. The kind of compression on which Canada's Telidon was based presumes the use of discrete, differentiated symbol sets, defined in advance by explicit coding protocols. As Licklider points out, this "mode of experience" is all about "the art of expressing ideas through configurations and manipulations" of such signs and symbols.[65]

Telidon, a second-generation videotext format and technology developed between 1976 and 1985, serves here as object lesson for a specific contribution to the media history of compression. Telidon is generally remembered today as one of a class of teletext and videotext systems, such as the much better-known French Minitel, that are typically accounted for as precursors to the internet. Telidon resonates still as a major financial fiasco for its Canadian sponsors, or within media arts history as the site of experimental uses by artists that were till recent years murkily documented.

The underlying technical work leading to the Telidon project, an initiative of the Canadian Department of Communications, occurred in the early to mid-1970s. It was carried out at the Communication Research Centre (CRC) in Ottawa. The CRC was, and remains today, a departmental branch laboratory dedicated to applied rather than basic research. Like so many institutions that would become associated with pioneering work in computation and networking, the CRC contributed to classified signals intelligence during World War II, and in the 1960s to the Canadian satellite program. The lead inventor, Herbert Bown, had developed hardware designs and software in the early 1960s to support sharing of graphic designs of electronic components, for which he devised a scheme for efficient coding of picture descriptions that could be conveyed over low-bandwidth transmission. The initial inventive concept appears to have been that of a common visual space, in which multiple users could view and update a shared graphic display from remote screens, while conducting telephone voice conversations (see figure 6.2).[66]

The hand-held pencils writing on the sheet representing a common visual space are in effect "processing units" with the capacity to decode and render messages conveyed over the telephone wires. The specific application domain was initially intended to be collaborative circuit design. This early in-house research project attracted the interest of Bernard Ostry, a senior bureaucrat in the Department of Communications, who had been monitoring developments in the British Prestel teletext system. Based on

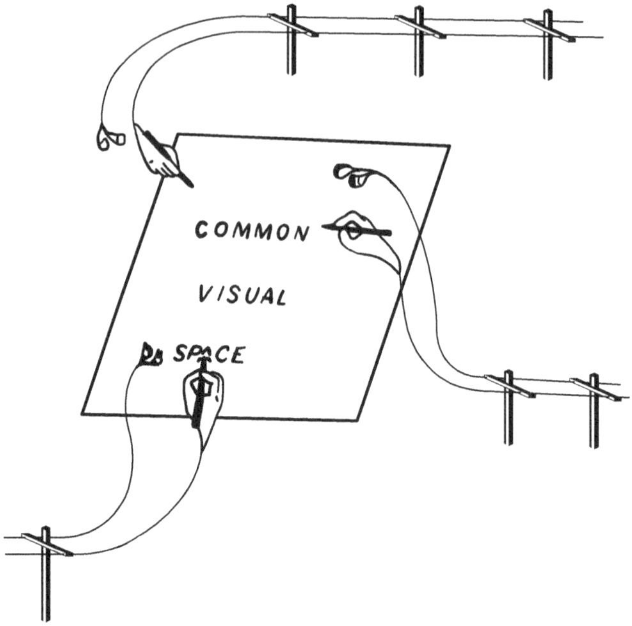

Figure 6.2
Concept of a common visual space. Adapted from Sawchuk et al., "An Interactive Image Communication System Using Narrowband Lines," *Computers and Graphics* 3 (1978).

what he saw as the wider communicative potential inherent in Bown's concept, Ostry charged him and his group with the task of imitating and improving on Prestel, for Canadian and eventually export markets. The chief improvement was in the expansion of the picture description instruction scheme that would ultimately become Telidon's signature feature and selling point: a second-generation videotext system offering alphageometric, rather than alphanumeric, coding.

In 1978, almost immediately after this initial work had been completed, and well before the Telidon program's official launch, Bown and a colleague from the CRC, William Sawchuk, found an important practical application for their method of compressing meaning through picture description instructions at a Toronto center for disabled children. The clinic, then called the Ontario Centre for Crippled Children (OCCC), treated nonspeaking patients afflicted with cerebral palsy. In 1971, the OCCC adopted the ideographic communication system developed by Charles Bliss, as outlined in his book on "semantography."[67] Bliss symbols can be understood in the long history of invented philosophical and universal languages, such as those designed or imagined by Leibniz, Comenius, Campanella, and Wilkins.[68] Bliss was an Austrian engineer who, after surviving the Nazi concentration camps during World War II, migrated to Shanghai and observed the use of ideograms in the Chinese language. He designed his symbol set with the intent that it serve as a universal pictorial language—a "direct line to concepts."[69] The OCCC found the Bliss symbols to be effective in allowing the disabled children to communicate sophisticated ideas, not just to select from a preexisting set but also to create new "words" by combining them in original ways.[70] Bown and Sawchuck adapted the CRC system to display the Bliss symbol set, and after some trials in use with the children, they were able to report:

> If the problem of designing particular interfaces for physically-handicapped children can be overcome, then the effects of an interactive graphics system to symbol communication can be profound. Such a system employed over existing telephone lines would certainly broaden the scope of the children's communication environment. An at-home terminal would readily enable them to communicate with their friends, relatives and therapists. . . . This use [of symbol communication] need not be restricted to physically-handicapped children. Symbol communication may be applicable to persons of all ages and to other types of disabilities.[71]

A similar view was expressed by Canadian Gordon Thompson in 1983 when he extrapolated from the static Bliss symbols, as used by children at the OCCC, to advocate for the potential of a "dynamic, or time-varying iconic form of written language" using computer graphics.[72] Thompson's ideas will come back in a later discussion of the uses of the deployed Telidon system itself.

The official Telidon program was launched in 1978, funded by a series of departmental allocations along with industrial contributions that in the end topped $100 million. Deployment involved several parallel phases, including trial services offered by telephone and cable companies, information providers like newspapers and broadcasters, as well as libraries, schools, and universities. Hardware manufacturing of decoding modules and terminals was also subsidized, aiming to kickstart a competitive Canadian digital display and component manufacturing industry.

In understanding the Telidon rollout, it is crucial to outline the governmental backdrop against which all this extensive intervention occurred.

Until 1979, all cultural affairs funding and policy at the federal level were handled by the Department of the Secretary of State. Only in 1979 was this policy area merged with the Department of Communications, which had been set up in 1969 and handled telecommunications and communication research, as well as content regulation. This merger of culture and technology, content and carrier, or as the idea was sometimes termed in Ottawa, of software and hardware, occurred early in the attempt to boost demand for and public enthusiasm for Telidon. By then the program could be rhetorically lionized by the first Liberal minister of the combined department, Francis Fox, in 1980:

> It is undeniable that new technologies have the power to atomize us, to make us isolated individuals individually connected to a multimedia computer center. But they also have the power to come closer, to give birth to the global village, or rather to the global city of which Telidon would be the public place, the town crier and the storyteller.[73]

Similarly, in a promotional film made for the Department of Communications by the National Film Board, David Suzuki, already Canada's preeminent science communicator, touted Telidon as the latest breakthrough in humankind's ability to tell stories, record ideas, and pass down culture from generation to generation. Suzuki introduced Telidon to Canadians as the "new vehicle for the computer age; a space-age medium that combines the eternal force of the printed word, the speed of the computer, and the globe spanning web of modern communication."[74] Implying that these previous media would be not only absorbed but surpassed, Suzuki underlined that Telidon is "immune to obsolescence," because it is built on content standards that would be adaptable to future changes in hardware. The film then goes on to picture users engaged in now familiar (but then nascent) transactional online services: shopping, real estate management, telebanking, travel booking, and educational games.[75]

The push to establish Telidon technology as an international, not just Canadian standard was carried out in conjunction with US telecom giant AT&T. The resulting North American Presentation Level Protocol Standard (NAPLPS) was to figure as the primary selling point of the technology export push by Telidon's champions, along with resolution scalability, device independence, and low latency.

In a series of articles published over three issues in 1983, the popular American home computer hobbyist magazine *Byte* presented NAPLPS as an advanced standard for digital communication designed to allow a large variety of information to be encoded in a manner that "preserves the conceptual content of the information" across diverse rendering resolutions, device types, and transmission rates.[76] It is interesting to note that to the *Byte* magazine author, the idea of semantic conservation might be an answer to an unposed problem, since "most are trying to figure out what this [conceptual conservation] means and why it is useful."[77] As it turned out, the idea never caught on with the broad public: even at Telidon's height, and despite the considerable financial and promotional support, no more than 10,000 Telidon terminals or kiosks were in use. The government finally canceled the program in 1985.

Previous scholarship on Telidon is of three broad types. Communication and media scholars writing near the time of the program aimed to analyze the system's failure to achieve widespread adoption. These early accounts were critical of the technology push—rather than the demand pull—at the root of the project.[78] Lacking attractive information services, users had little incentive to enroll by purchasing relatively costly decoders and terminals.[79] Hostile early critics of the postindustrial society attacked the "pushbutton fantasies" of videotext cloaked in the typical period hype of dream machines, creative participation, and telework; one such critic, Vincent Mosco, softened the venom of his attack against Telidon, compared to the commercial US initiatives, owing to its having set up a consultative committee.[80] On a more technical level, a second broad type of criticism focused on Telidon's failure to overcome a basic confusion between its aspiration to become an international content standard and its potential to underwrite new kinds of born-digital sociability. Elaborating on this criticism, in comparison with the French Minitel system—also state-mandated but with lower cost to users—Jean-Claude Guédon implied that Telidon failed to take off because it did not afford users with a genuine zone for social experimentation.[81] This was in contrast to what Andrew Feenberg called the "social construction" of Minitel, which came about when French hackers were able to open the system up to third-party creative services, initially through the *messageries roses* enabling anonymity and playful amorous encounters.[82] Third, Telidon has been criticized for concentrating mainly on the worldwide acceptance of the NAPLPS protocol and ignoring the development of the personal computer, graphical user interfacing, and data communication standards (like the internet's TCP/IP), as distinct from Telidon's presentation layer standards.[83]

Less critical accounts have celebrated the system on various counts. At the Communication Research Centre itself, Telidon figures as a research milestone for having contributed to the development of international standards for the World Wide Web. In a 2006 article in the *Annals of the History of Computing*, videotext systems, including Telidon, are credited with taking "pioneering risk" in advance of internet commercial services' viability.[84] More recently, arguments have been made for the valuable role played by videotext systems like Telidon in establishing the success of the web by developing a base of users who were cognizant of the potential value of online services. This capacity-building argument is made also by some in the media arts community, notably by analysts of the Telidon artist trials. These trials enabled artists to gain basic familiarity with graphics computing, broad programming literacy, and hypertext logics.[85] Canadian communication scholar Kim Sawchuk introduced a theme issue of the *Canadian Journal of Communications* on experimental practices in the media arts with a reference to these trials as "legendary yet under-studied."[86] Influential figures in computer graphics education and design research recall the Telidon system in terms of capabilities development, for having challenged early users to see "how far could you go with so limited a tool": "artists used Telidon against its limitations . . . with the most primitive constraints, they used it as an animation system against its design."[87]

An example of just such a counterintuitive deployment of Telidon as (at best) a clunky animation system actually appeared in another NFB film, commissioned by the Department of Communications with the apparent intent of demonstrating the system's flexibility. *Graphic Variations on Telidon* was directed by Pierre Moretti, and consists of a series of exercises to demonstrate the dynamic and color graphics capabilities of the system. Nested expanding circles pass into rotating moiré patterns and cumulative patterning of overlapping rectangles, triangles, and trapezoids, concluding with a demonstration of quasi-3D blocks and a set of jittery oscillating triangular "creatures" moving awkwardly across the screen. Accompanied by a portentous synthesized electronic music score, Moretti's *Graphic Variations on Telidon* reveals that, indeed, the system's graphics mode makes simple and attractive geometric imagery possible, but the film did not receive much distribution; nor is it even included anymore in the NFB's extensive animation catalog.

A more insightful creative application of Telidon was developed at the University of Québec by Michel Cartier, who worked with Telidon as a research platform for *communication télématique*. His experiences are reflected in a text on design for "mediatics" that was included in a 1983 book promoting the system, edited by Canadian novelist and software entrepreneur David Godfrey.[88] Cartier well understands the basic challenge of compressing complex messages into concise visual messages. As a design theoretician, and in contrast to David Suzuki's recital of a progressive sequence of media advances all leading to Telidon, Cartier begins by pointing out that "books have not been the only types of records used down the centuries. Studying these older writing types suggests surfaces for telematic writing."[89] For Cartier, epigraphs, ancient Greek stone sculpted stelae, and ideogrammatic inscriptions offer earlier parallels where a reading of "inter-related glyphs [can] be understood." Cartier's understanding of the affordances of telematic compression led him to propose a new design discipline based on his original conception of a "modern epigraphics." He insisted that telematic design will have to become a time-based art, that is, one that treats reading as a temporal process:

> While books are spatially oriented . . . time is the essence of how telematics complements the written word. . . . The mediatics specialist activates different zones on the screen . . . [drawing] attention to a series of central ideas around which secondary ideas will be developed. The mosaic of structured information created through alphanumeric displays is the result of an underlying and appropriate use of time.[90]

Gordon Thompson, quoted earlier in connection with Licklider's encouragement for just such an ideogram-based form of representation, wrote extensively on the topic of time-varying icons while widening his own support for the kind of work Bown had done with nonspeaking children at the OCCC in Toronto. In a 1983 article addressing "Visual Literacy and the Time-Varying Icon," Thompson encouraged a dynamic writing language, "like American signing, adapted to a dynamic or time-varying form using computer graphics." He went on to argue that "conventional ideographic writing forms, being static, require a very large number of symbols and are hard to learn," instead encouraging the creation of new "techniques [to] reduce the effort required to assimilate information . . . in a world where mere abundance of information is a problem."[91] However, by 1983 his earlier hopes for Telidon's potential to provide the infrastructure for supplanting a classical view of literacy with dynamic time-varying icons were already dashed. "If videotext continues to be used as if it were an extension to the printing press, then we technologists, who really should appreciate the potential of our creation, have failed our brother citizens."[92]

In 1983, the same year Thompson expressed this disillusionment, the Toronto-based Telidon artist trials were launched. Initially these were carried out at Trinity Square Video, an artist-run production and access center, and later in a spinoff community videotext cooperative organization which would become Inter-Access, still one of the most active of the Canadian media art centers. A grant from the Department of Communications fund for Cultural Initiatives paid for the purchase of Telidon equipment. The project instigator was William Perry, who, in the catalog for an exhibition of the Telidon artists' works at A Space gallery in November 1985, provided an insightful media-theoretical overview of the system's potential as he saw it:

> The ultimate future is the complete digitization of any and all forms of representational art, from numbers, letters to pictorial representations; the most ambitious is the third. Telidon is a code that "compresses" video signals so that they can be sent over low bandwidth carriers. Theoretically, "video compression techniques" will progress from geometric primitives (rectangle, circle, arc, etc.) to full motion pictures over the ordinary telephone. . . . Simultaneous to this are developments in digital audio reproduction, sometimes referred to as Music Description Instructions (MDIs).[93]

With tongue firmly in cheek, Perry concluded that the Telidon coding protocol, NAPLPS, would eventually be the most significant updating of rhetoric and stylistics since Aristotle's *Poetics*, and urged learning its syntax as one would Strunk's *Elements of Style*.[94]

Several of the key participants in the artist trials recall the initiative in more cynical terms, as a deliberate attempt to subvert the corporate-government agenda of Telidon that was so firmly in place by the early 1980s.[95] Paul Petro, co-curator with Geoffrey Shea of the A Space exhibition, recalled:

> We knew what was on the horizon was banking and teleshopping, so we tabled the idea that the artists are the immunity system, from McLuhan: what would be the social and psychological consequences of the technological advances? To immunize—to take a humanistic approach, to appreciate and articulate the capacity of artists' use of technology.[96]

Petro's compilation videotape, *Canadian Artists in Teledon*, remains in circulation through the Toronto distributor V-Tape. It consists of twelve short graphics experiments, mainly rotating and twirling geometric abstractions, color-cycling, and slow-time fills and subtractions of dots and lines. Remarkably, Petro produced this videotape with government support, this time from the Department of External Affairs, for showcasing Telidon at the World Economic Forum in Davos.[97]

Toronto artist Nell Tenhaaf's two Telidon artworks, *Us and/or Them* (1983) and *Shopping for Value Systems* (1984), are principally engaged in critiquing the logic of hierarchical menu systems. The works effectively point out the limitations of preformatted choices in the much-hyped version of interactivity on offer at the time. In both, Tenhaaf interrogates preconfigured interactivity in a way that anticipates later critiques of the supposed democratic potential of new media, such as Alexei Shulgin's mockery of the banal manipulation of interactive systems.[98] *Us and/or Them* consists of fifty screen pages of text, maps, and graphics depicting the Cold War power struggle between the United States and Soviet Union. As Kim Sawchuk notes, the content and structure of the database "established an analogy between the binary structure of information based on a Boolean logic of 'and/or' and the deadlock of an us/them Cold War mentality."[99] *Shopping for Value Systems* similarly satirizes the commodification of choice in online shopping systems.[100]

Both Pierre Moretti's *Graphic Variations on Telidon* and the works produced by Canadian artists in the officially sponsored Telidon trials appear curiously tentative when compared to the varied ways in which teletext (the earlier, nongraphic version of videotext) was used by artists in other countries. When the Minitel system was licensed in Brazil, for example, a group of experimental poets used it to produce and display online concrete poetry, works that appear to have a striking effectiveness and formal coherence thirty years later.[101] In Paris, the Minitel terminals served as the main examples of computerized media functioning in Jean-François Lyotard's 1985 landmark exhibition exploring "immateriality" at the Centre Georges Pompidou. Roy Ascott, the British cybernetic artist seen in chapter 4 as president of the Ontario College of Art, and in this chapter as co-curator of the *Planetary Network*, had previously organized experiments in networked online writing that served as the inspiration for a significant component in Lyotard's *Les immatériaux*.[102] This component consisted of twenty-six authors contributing commentaries on fifty keywords, and illustrates well how the Minitel system, simpler and less technically advanced than Telidon, was able to support an exploration of the postmodern sensibility that was central to Lyotard's philosophical thinking about the uses of new telecommunication technology.

In comparison, Canada's experience of cultural networking potential using Telidon was almost wholly a disappointment. While a notional "right to communicate" circulated in government treatises,[103] and Communications ministers touted it as a boon to civil society against dehumanizing technologies, Telidon ultimately lacked "a people." The cultural uses of Telidon, when they weren't stereotypically transactional, were for the most part tentative or belated, and too furtive to permit a sustainable imaginary to emerge, one with the potential to challenge, provoke, and influence the assumptions underlying its compressive infrastructure.

Epilogue

By the mid-1980s, standalone microcomputers, operable with interfaces for nonexperts and stocked with application software marketed to enhance creativity in the arts, were quickly sweeping into widespread use. Dial-up access to online services was taking off, while the noncommercial internet, still government-subsidized, was increasingly accessible outside academic circles. As Robert Adrian observed, the heady, anarchic "pioneer" period of globally oriented artist networking using hybrid systems—fax, print, phone, slow-scan, email—had peaked,[104] while satellite-based video connectivity was becoming more and more commonplace. Telepresence moved from situated performance to commonplace use, whether in the soon-to-be-typical sight of multiscreen live talking heads on commercial broadcast; in bespoke institutions, like the Los Angeles-based artist hangout Electronic Café equipped with video conferencing gear; or in emblematic special extravaganzas, like video artist Nam June Paik's 1984 New Year's Day multi-simulcast, *Good Morning Mr. Orwell*. On the Canadian experimental scene, artist-led centers began to specialize in emerging digital practices that would, by the 1990s, come to be lumped together as "new media art"—first the Artculture Resource Centre (1981), then InterAccess (1987), both in Toronto, and later the feminist Studio XX (1996) in Montréal, being indicative. Meanwhile, the network of artist-run centers as a whole was in flux, less driven at large by media experimental concerns and more by issues of ethnicity, gender, and racial inclusion and social justice.[105]

For Tom Sherman, founder of the Canada Council's Media Arts Section, the blending of actual physical spaces and communications networks in Canada's artist-run scene had been both instrumental and prescient. He compared the Council's and other Canadian funding policies with those of other countries at the time:

> France and Germany were stockpiling their media arts resources into organizations like IRCAM [Institut de Recherche et Coordination en Acoustique et Musique] in Paris and ZKM [Zentrum für Kunst und Medientechnologie] in Karlsruhe. Against the logic of this international technological arms race, it was also becoming obvious that technology itself would be soon decentralizing. . . . Personal computers were on the immediate horizon. Technology itself was also a force and digital telecommunications networks were exploding in the early 1980s. . . . Artists were breathing new life into traditional forms and skirting art forms entirely by speaking directly through proliferating communications channels.[106]

This chapter has set in counterpoint two faces of cultural networking, one based on the freshness and novelty of real-time mediated engagement, the other on the variously understood affordances offered by standards-based content compression. The playful use of mediated engagement as an art form, where "the user is the content," would become commonplace, after the internet was widely adopted after 1992 and evolved into the business model of today's corporate social networks where the user quite literally *is* the product. The narrative of verisimilitude, understood in terms of the density of display resolutions and the multiplication of transmission rates from hundreds of bits to gigabytes per second, would make mainstream considerations of the expressive intensity of compressive aesthetics seem archaic.[107] Synthetic pictures were becoming photorealistic, sonic as well as graphics environments were becoming immersive, and the interface was starting to be conceived in holistic terms as the entire incorporation of the senses and the inclusion of fully embodied expression. The horizon for this next round of the "technological arms race" was coming into view by the late 1980s, and in one instance in Canada it congealed around virtual reality, at once heralded and contested as a new artistic medium.

7 Contesting a "New Art Form": Virtual Reality as Cultural Probe

The central theme of this book is the manifold ways that the alternative technological ethos in Canada stimulated artistically driven approaches to new media forms at an early developmental stage, prior to their stabilization and commodification following wide adoption. Exploring the boundaries of media interface and sensorial experience, the alternative ethos was pointedly invoked within a prolonged exploration of virtual reality during its earliest phase of mass media hype. This central theme is also inextricably bound up with Canadian government policy priorities hoping to instrumentalize the supposed creativity of the experimental artist in order to seed new homegrown information enterprises through directed funding. The instability of this complicated formula, combining research-creation and discursive critique with economic and technology development interests, played out at an unlikely location, itself somewhat mythologizing: the Banff Centre for the Arts, nestled on a mountainside campus in Canada's first national park.

The Banff School of Fine Arts[1] was founded in 1933 during the height of the Depression as a summer school with a mission to support a "theater of the people." During these early years, it offered drama workshops for ordinary prairie folk to explore the challenges of their lives and dreams during that period of want and turmoil.[2] Over the following decades, it became a classical arts mecca—sometimes called the Salzburg of the Rockies—a busy conference center and management school, and a global hub for education about mountain culture. Internationalist in spirit, but by mandate also locally grounded, the Centre's programs were supported financially by the provincial government of Alberta, which accorded it a quasi-university status while leaving it free to offer flexible programs of varying lengths, without needing to be accredited as a degree-granting institution. A year-round postgraduate program in the visual, performing, and literary arts was established by 1980. Contemporary art exhibitions interrogated the tourist economy of the national park, its colonialist pasts, and the paradoxes of a contemporary art center operating alongside a school for executive business administration. The Centre also launched programs in contemporary musical theater, sponsored a three-week workshop in "improvisational composition," and put in place a boundary-bending organizational scheme for the promotion of "inter-arts."

The philosophy of the Inter-arts program had been set out in a 1978 position paper that urged the Banff Centre to become a "laboratory" for the creation of new work across the arts, not just open to but deliberately encouraging experimental collaboration between fields.[3] What began as an emphasis on exploring connections between the established arts disciplines expanded by the middle of the next decade into a practical interest in the role of new communication and media technologies in catalyzing and enabling new forms of interdisciplinary creation. Visual arts curators Daina Augaitis, Barbara Fischer, and Lorne Falk organized exhibits and artist residencies around topics such as electronic visualization, "rhetoric, utopia and technology," "siting technology," and radiophonic and sound art.[4] Under my direction, the Inter-arts program opened a small computer lab, then spun off a media arts program in 1988 with a hybrid research and production mandate. Its founding manifesto anticipated a range of new or transformed art forms that would be the focus of this mandate:[5] hypermedia, electronic opera, interactive music and immersive sound art, and what was initially called an investigation of "virtual worlds." The last of these—the investigation of virtual worlds—would emerge as a principal focus for the whole program by the early 1990s, as the first wave of hype and investment in VR spread into the arts from the military, architectural design, and space-exploration-focused "virtual environments" labs located in Silicon Valley, Boston, North Carolina, and Seattle.

This chapter analyzes a sustained effort to *contest* virtual reality's emergence as a putative "new medium." Here the alternative technological ethos playing out across the chapters of this book was explicitly framed as a critical resource to challenge and deflect mainstream technological trajectories for VR. The word "contest" is used here in a sense derived from the Greek word *agon*, following the incisive theorization of interdisciplinarity advanced by

Andrew Barry and Georgina Born.[6] Examined today, the contest over the meaning as well as practical build-out of virtual reality that took place at the Banff Centre between 1989 and 1994 appears as a process drawing together several distinct modes and logics. Technological innovations and economic benefits were central to the interests of funders in the Government of Canada, as well as among the corporations contributing resources and personnel. Utopian-tinged syntheses of art and technology were imagined by some of the research participants, who were mainly, though not exclusively, technologists. Critics, artists, and philosophers, particularly those coming from feminist perspectives, sought to resist and subvert the ideologies of disembodiment, dematerialization, and universalism they perceived around the "mythology," if not yet the actuality, of the nascent state of development in VR at the time. Some of the organizers, including myself, also envisaged a durable research policy advance, whereby the definition of VR as a "new medium" would be shaped by feedback loops fed both by artist-led prototyping of new kinds of objects and radical cultural critique. Altogether, this commingling, or "mangle,"[7] of modes and logics for interdisciplinary research can also be set in wider perspective still, positioned against the backdrop of Canadian communication theory. The trope of the artist as cultural probe—the primary person, according to Marshall McLuhan, capable of critically recognizing both the perils and the promise of new media technologies—permeated the entire Banff enterprise in VR.

The Banff Centre as Studio Laboratory

When the Media Arts program was put forward to the Alberta government's Department of Advanced Education as part of the Banff Centre's 1989 operating plan, the driving ambition was "to provide an environment in which the artistic, social and cultural significance of media technologies can be explored by artists working in all disciplines."[8] Academic partnerships were established with Alberta's comprehensive research universities in Edmonton and Calgary, and the computer science departments at both were involved with the Banff artistic and technical teams from the start in joint seminars, exchanges, and by offering internships at Banff for their graduate students. These university communities provided a pool of scientific and engineering expertise in the initial stage, though links were also established from the outset with the National Research Council in Ottawa, the province's own Alberta Research Council, NASA's Ames Research Center in Mountain View, California, as well as several prominent software and computer companies. For three-dimensional computer graphics software, Banff obtained contributions from Alias Research, a Toronto-based firm whose founding president was intrigued by the rising buzz around cyberspace and virtual reality. Mark Green in Edmonton directed one of the earliest research labs for virtual reality in Canada, with high-performance graphics workstations and head-mounted displays. Green had done his graduate work at the University of Toronto, focusing on human-computer interaction, and had worked closely with Bill Buxton on computer music there as well. Catherine Richards, an artist then working as a consultant for advanced imaging at the Canadian Broadcasting Corporation, contributed the initial conceptual foundation needed to formulate the Banff Centre's multistaged research program oriented primarily around artistic experimentation informed by critical theory.

In 1987, Richards had assembled a group with this kind of intellectual range for the American Film Institute in Los Angeles, in a panel that included Scott Fisher, the lead researcher on virtual environments technology at the NASA Ames research lab, Canadian political scientist Arthur Kroker, and Ann Marion from Apple. Richards contextualized VR for the Los Angeles film community in terms of unprecedented representational technologies, explaining that the camera would have a point of view within the scene: "Not only can this camera (the user) direct its own looking, it can be sensed, responded to, and represented in the scene: it sees and is seen." Elaborating on the epochal significance of this fact, she went on to set the technology within a particular genealogy of the avant-garde arts:

> We have passed through a period of art which included performance work, theatrical improvisation, happenings, "guerrilla" TV, social action and community video, where much of the emphasis was on participation, interaction and the formidable spectator/object boundary. These are characteristics of the new media, where the viewer/spectator/participant/user and object/machine environment are in close embrace. *It is their close interaction that creates virtual events.* Digital images are only a part of this virtual world's new suspension of disbelief.[9]

Also in 1987, at a very different venue—Arthur and Marilouise Kroker's arch-postmodern edited volume on *Body Invaders*—Richards published a graphics-and-text artwork entitled *Body in Ruins*. The piece collaged NASA's already iconic drawing of an eyephone-helmeted operator in a robotic telepresence application with a textual layout of the words "body in ruins," and "we—are—wearing—it—out."[10] Richards's interest in interrogating the permeability of body boundaries—"what it means to go so close to the senses"—was foundational for defining what would eventually be termed the "Bioapparatus,"[11] a complex, sharply contested conception that would serve as the basis for Banff's first foray into virtuality.

Richards can therefore be seen as a strategic boundary-spanning individual, capable of doing the translation work needed to bring ideas from science and university labs into cultural settings, while also understanding well the specifically Canadian policy context in which such information flows would take place.[12] Trained in literature and fine art, Richards was employed in the 1970s in the research department of the Canadian Radio and Television Commission. There she was immersed in both the intellectual and technological ferment of Canadian communications discourses within a state institution with large responsibilities for telecommunications at large, responsibilities which included long-range technocultural forecasting (as in the *Instant World* report discussed in chapter 1). The leader of Richards's research unit was André Martin, the French film critic who had come to Canada in the 1960s primarily because of what he saw as Canadian vitality in media theory and film innovation.[13] (He began his time in Canada making a film for the NFB on McLuhan; see chapter 2 for discussion of Martin's writings on Norman McLaren's "instrumental cinema.") In Richards's recollection of those years, Martin ran what was in effect a think tank on communication, not a conventional audience research shop, and the scope of the work allowed her to visit and interpret a wide range of developments in video art, community media, and computer images and interactive techniques. In an atmosphere fed by the ideas of Canadian intellectual heavyweights like Harold Innis, Northrop Frye, and McLuhan, on one hand, and the research at the National Research Council, National Film Board, and the University of Toronto Computer Science Department, on the other, Richards viewed her role in specifically McLuhanist terms, as a human "probe" for interrogating the potential of new media environments. Just as Kar Liang had immersed himself in film production while working as an engineer for the National Film Board's technical research unit (chapter 2), so too, from the other direction, Richards was able to gain a firsthand view of the US laboratories over this decade of working in graphics and interactive techniques.[14]

As a visitor to the NASA Ames research lab and other corporate and government locations, Richards formed professional contacts with researchers and insights about the techniques of immersive real-time visual and sonic environments. Her perspective as a "probing artist" was to look at the early emergence of VR, before the wave of hype that began to build around it in the late 1980s. VR, as she thought of it then, could serve as a "template" for asking questions about how this kind of intimacy between the human body and immersive technologies would develop.

Alongside Richards's work during this time, Banff Centre curator and critic Lorne Falk directed the Centre's studio art program, which was organized as a series of professional-level residencies around thematically organized topics. Falk formulated an artist residency complementary to the one Richards was developing around virtuality, working with consultant Nell Tenhaaf. Tenhaaf's contribution was initially defined in terms of "boundary degeneration" in the context of *both* biotechnology and virtual worlds. As articulated by Tenhaaf, the underlying thinking had a marked feminist orientation, with the aim of understanding "how knowledge is engendered, and gender-determined, through androcentric scientific thought. Under conditions of the historical devaluation of women's thinking, how we know is not gender neutral."[15] Combining the interests of media and studio art, oriented respectively around the inputs of Richards and Tenhaaf, the collaboration expanded to include the Centre's Walter Phillips Gallery, through its publishing program.

Having compiled these various ideas and plans and submitted them as a funding proposal to the federal Department of Communications, I was invited to deliver a keynote speech to a national conference on culture and technology in Montréal that the Department was sponsoring. Along with Martin Tuori, head of research at Alias, we took the occasion to outline the "stakes" in VR from a combined technological and cultural-theoretical standpoint. Tuori demonstrated the graphics capability of his firm and provided a technical grounding in the hardware and software functions, while I used the platform to pose the rhetorical question—"as real as what?"—while outlining Banff's VR project and acknowledging the contribution of its initial funder, the short-lived Canadian Institute for Research on Cultural Enterprises (CIRCE). The following excerpts indicate how the Banff Centre's undertakings were being publicly presented at the time.

> When we ask "as real as what?", we are drawing attention to one among many curious facets of the current virtual reality phenomenon, namely the huge gap between the claims made by its proponents and the present capability of the technology. In short, we cannot avoid the fact that already, in the few short years since the early prototypical systems were launched as commercial products, a cult of virtual reality has been born, complete with hucksters and hype-masters heralding the arrival of yet another new age of personal empowerment through media-induced consciousness expansion.

After identifying the mythology around the fetishized VR of the early 1990s, the speech went on to outline some of the aesthetic questions emerging out of Banff's proposed VR "probe," beginning with the then hot-button postmodern issue of simulation.

> We have already mentioned the way in which artists have refused to accept the doctrine of neutrality regarding technology—that tools are indifferent to the purposes or creative intent brought to them. It seems a necessity, at least in the introductory stage, to challenge the assumptions of technology designers, especially where these assumptions might be gender-biased. . . . The philosophical issues that we think require research by artists are indeed profound: Questions about representation, abstraction and identity; photo-realism: how much reality is enough? Industrial or corporate applications may have an insatiable need for visual resolution, but what bandwidth is sufficient for plausible interaction in a virtual world? How to signify a tree? With leaves and shading, or just an icon? Personal representation? How important are other senses? What do we lose by not being there? Which of the laws of time and space do we replicate? Does virtual reality need gravity?[16]

Addressing the artists in the audience directly, we acknowledged that while "there may be suspicion on the part of artists about the technologies of control in virtual reality systems, a good number are well primed, from the conceptual standpoint, to take creative advantage of the technology as long as they can approach it as a true artistic milieu. What in fact is required to effectively place a new or recently emerged technology into the hands of artists?"

At the conference, the Department of Communications announced a $50,000 startup grant to support the project's first phase, a ten-week residency on "the Bioapparatus." The grant was approved in the midst of a controversy over the botched launch of a new culture and technological institute within the Department of Communications: CIRCE had been announced earlier in 1991 by the Department's flamboyant Québecois minister, Marcel

Masse, with a fifty-million-dollar budget, and was slated to be headquartered in Montréal. An ambivalent federalist who affected the airs of a European-style cultural czar, Masse had worked closely with David Godfrey, an award-winning novelist with a background in high-tech communication (including Telidon marketing in the cultural sector) to define the government mandate for CIRCE. The idea was to combine policy and statistical studies on what would now be called the "creative industries" with applied research in the new media and communication technologies themselves. However, the fledgling organization was soon caught in the crosshairs of Canada-Québec political controversies, which in those years turned on the failed efforts, during the late 1980s, of Prime Minister Brian Mulroney to bring Québec into Canada's Constitution with the special status (rankling to many anglophone Canadians) of a "distinct society." Anglophone commentators saw Masse's announcement as a blatant patronage bauble to Québec and ridiculed the public's and journalists' erroneous impression of its mission as limited to gathering data on cultural industries and institutions. The policy and observatory component of the CIRCE plan was receiving all the publicity, while the institute's other intended focus on new technology research was drowned out by the political furor. Godfrey tried to counter some of the press attacks in a set of talking points faxed around the country:

> No area of the arts is exempt from new technologies and advanced economies such as Germany and Japan are spending hundreds of millions of dollars to be prepared for the future. For Canada, 4.5 million per year seems like a minimum investment. . . . Half of the funds of the institute have been dedicated to providing Canadian artists and enterprises with better access to knowledge and tools to help ensure that we are preparing for the future and not the past. To do this in a cost-effective manner, the Institute will act as a coordinating center and form affiliations with other existing sites of excellence across the country. These, in turn, will act as centers for other groups within their region and/or area of excellence. The whole will be linked via networks, shared databases, and electronic conferencing.[17]

Thus was the Banff Centre's Bioapparatus funding from the federal government justified and announced in the spring of 1991, just in time to organize the program's ten-week residency between October and December of the same year. By June, CIRCE had been shut down, and Masse was sacked as minister of communication, though not before securing a further half-million-dollar grant for the second phase of Banff's initiative, a follow-up commissioning and production phase intended to grow both conceptually as well as technically out of the more exploratory interdisciplinary residency. The overall project, as approved, would "establish an artistically-defined laboratory" for virtual reality, in conjunction with academic and industrial collaborators, with the mandate "to research and experiment with virtual technologies as artistic media." Its multifarious projected benefits were to "establish Canada as a significant player in virtual reality R&D; contribute to the growth of Canadian high-tech infrastructure through pioneering industrial-academic-artistic synergy; create the first significant works of art in virtual reality; enrich critical discourse on art and technology; communicate the above to the public and scientific community through documentation and public events."[18] Following the ten-week exploratory residency in the fall of 1991, an international call for proposals was issued inviting artists and groups to submit their ideas for a subsequent full-scale production phase of VR artworks. By spring 1992, a dedicated laboratory was put in place at Banff with state-of-the-art graphic workstations and peripherals. Technical staff were recruited mainly from Alberta computer science departments, and eight full production projects were selected. The completed works were exhibited in May 1994, in the context of an international conference; a book documenting these final projects, with commissioned essays, was published in 1996 by the MIT Press.[19]

Agonies of the Bioapparatus

The roster of the twenty-five Bioapparatus participants was determined by invitation and was put together collaboratively by Richards and Tenhaaf with the Banff media arts and visual arts staff. An attempt was made to achieve a balance of gender, medium (in practice, divided between sound and image), and role (artist, technologist, and theorist/writer). Multiple intersecting identities were commonplace.

- Mary Ann Amacher, composer, performer, multimedia artist, New York, known for architectural staging of music and sound with a foundational knowledge of psychoacoustics and electronics.
- Wende Bartley, electroacoustic composer from Toronto, concerned to create sonic landscapes that help the listener access knowledge about the collective story of woman.

- Eleanor Bond, painter from Winnipeg, whose images are visual forms of future fictions.
- Diana Burgoyne, Montréal, whose performance work uses the body as a transmitter to explore the relationships between people and technology.
- He Gong, Chongqing, one of the principal activists of avant-garde art in Western China, in work that addresses postmodern culture and performance art.
- Doug Hall, San Francisco, was a member of Ant Farm (1960s) and one of the pioneers of video art.
- Carl Loeffler, founder of *Art Com*, an art-telecommunication center in San Francisco.
- Robin Minard, Berlin-based Canadian composer developing interactive systems which translate spectral analysis into construction of environmental-dependent music.
- Michael Naimark, San Francisco, artist-technologist straddling the research and art communities, whose previous credits included the MIT Aspen Movie-Map Project.
- Warren Robinett, designer of interactive computer graphics, commercial video game designer and researcher for virtual reality system at NASA Ames.
- Kathleen Rogers, London based video, installation and computer-based artist.
- David Rokeby, Toronto, inventor of the *Very Nervous System* for "aural virtual reality," whose uses include therapeutic applications for the disabled.
- David Rothenberg, musician, composer and writer, whose work examines the relation between technology and nature in the history of philosophy.
- Daniel Scheidt, Vancouver, composer and computer scientist working in electroacoustics and interactive sound sculpture.
- David Tomas, artist, theorist, and anthropologist from Montréal, whose installations explore image technologies from photography to computers and their effects on representation.
- Inez van der Spec, Dutch writer and feminist theoretician, writing on science fiction's images of cyborgs, alien bodies, and transcendence.
- Lawrence Paul Yuxeluptun, indigenous artist from Fort St. James, British Columbia whose painting addresses environmental and political aspects of native people's experience.

Published accounts of the residency during the 1990s were primarily focused on the two-day seminar that involved all the residents, chaired by Richards and Tenhaaf and organized topically around a series of one-page position papers assembled by invitation from the residents as well as from corresponding authors around the world. Among the several dozen remote authors who contributed texts were philosopher Jean-François Lyotard, psychoanalyst and art critic Jeanne Randolph, University of Toronto McLuhan Centre director Derrick de Kerckhove, artists Char Davies and Rafael Lozano-Hemmer, curators and critics Florian Rötzer, Itsuo Sakane, and Regina Cornwell, and Ars Electronic Festival curator Gottfried Hattinger.[20]

An indication of the breadth of this group's "construction of virtual reality" can be seen by the following list of phrases scattered through the seminar booklet, each separately completing the statement "Virtual Reality is ___":

> our most recent form of representational technology; a symptom; an instrument re-ordering perception and power relations; the new and easy armchair way to have an experience; one more effort to buttress the masculine ego against its perceived foes; the realization of the techno-mythological concept of the automaton; a fast way of going beyond one's limits; a theatre that contains the ideas of all things except, of course, the idea of itself; an oxymoron; just one big Walkman; the most advanced from of interactivity to date; a form of integration of the senses; an antidote to problems we have not solved in the world; the solution looking for a problem; a Bioapparatus.[21]

Several commentaries on the seminar stress the fissures and gaps that separated the participants, underlining unresolved arguments about gender, technology, and the politics of design. David Tomas diagnosed the seminar as follows:

> After two days of presentations, discussions and the behind the scene ruminations and arguments, including the articulation of frustration and despair on the part of a large portion of the women artists and theorists, there seems to be only one conclusion available for retrospective contemplation. While the majority of artists appear to have been theoretically and practically ill-equipped to deal with this new technology at the level of its technical organization, those involved in developing its hardware and software are equally ill-equipped to deal with its social and cultural dimensions as well as its political implications.[22]

Echoing Tomas's bitter assessment, Robinett and Naimark reported on the seminar in a 1992 issue of *Presence*, a scientific journal on VR and telepresence technology: "there seemed to be a conflict of agendas and a problem of not understanding one another's languages between two main groups. Art theorists and feminists seemed to speak one language and technologists another." Yet they acknowledged that the groups overlapped, that some didn't fit either category, and that the hot-house atmosphere of the seminar "did serve to get the residents talking to one another."[23] The *Presence* article concludes by listing the completed artworks in VR that were produced during the ten-week residency, one of which, by Yuxweluptun, is described below. One of the computer scientists, who would go on to be closely involved in software development during the subsequent production phase, provided a sense of how the controversies characteristic of the seminar were articulated:

> The most intriguing point was the issue of "gendered technology." I'm frankly doubtful that this makes sense, but it's not yet resolved in my mind. The debate seems to beg for empirical backing of some kind . . . I got a clearer idea of what issues are of concern to artists. Some artists are mostly concerned with the *idea* of technology and not how it works. I'm not sure they are interested in collaboration. On the other hand, for those who are interested in "building things" there is an evident language/culture gap that must be bridged if either party is to achieve their respective goals.[24]

For an experienced sound artist, visiting the residency just for the two-day seminar, the premise itself was too vague: "the body/technology interface is an issue that is so broad, so general, that artists seemed to be at the limit of their 'language competence.'" One of the most vocal of the feminist critics during the seminar judged that its positive value lay in the fact that "the Western tendency to universalize 'the body' was foregrounded so clearly that it gave me a clearer position on breaking that thought pattern." Similarly, for a skilled programmer-artist, the point of provocation was in revealing new insight into "what the body is, how much what we already call the body is interface. The body has for most of us been a basic assumption of our existence. . . . The seminar was particularly helpful in clarifying the issues that cause the most confusion when artist and technologist meet. In particular, the first-hand experience of failures in communication due to specialized languages was extremely valuable."[25]

Beyond the fissures around gender and social role (artist/technologist/theorist) that emerged out of the seminar, looking back at the Bioapparatus residency and symposium as a whole reveals the distinct differences regarding virtuality that could be discerned between visual and sonic artists. Mary Ann Amacher, for instance, urged her colleagues to think of life itself as "wholly bio-engineered," supporting Lyotard's paper, which forecast the long-term emancipation of life from the constraints of biological systems.[26] In another intervention that transgressed some of the stereotypical gender identifications of the seminar, Amacher spoke with approval about her desire to be able to exchange body systems with others:

> We could enter other worlds and other people's sensibilities that way. If you had such a thing as a perfectly tuned body, all your neurons and all your sensibilities, would you want to sell it for a simulation? . . . I would like an agent or helper in my studio that would do a lot of listening for me. Actually, I never thought of giving this agent a gender, male or female. I would like this new perceiver, one day to be a man, another a woman.[27]

In his presentation to the seminar, Robin Minard insisted that sound artists were already engaging the whole body, pointing to sound as the "primal sense": "Even in deafness I will feel its vibration. There is no subject or object in the world of sound. Because sound inhabits me, because it requires my body to resonate within, because it finds my resonances in specific and individual ways, I become, in a sense, the object." Minard observed that among some of his colleagues, there was an understandable but misguided concern about the unforeseeable dangers of allowing the transgression of bodily membranes: "Within the domain of sonic art, we have been playing with the inside of the body in a direct way since the beginning—this is what we do. We get inside people's bodies *physically*. . . . As a sound artist, I accept these openings into the body as the only vehicle of my expression." Daniel Scheidt underlined that as a composer working with interactive music systems, he had no choice but to have a deep respect for the performer who necessarily enjoys an intimate closeness to technology, understood as "an extension of my being into other planes of existence, other realities."[28]

Another music-sensitive participant in the residency was William Leler, a computer scientist who had earlier contributed a position paper incorporated into the project's funding proposal. The paper, also delivered at the Second Conference on Cyberspace in Santa Cruz, California, dealt with "actor-based simulation" in virtual environments.[29] Leler's critique of VR was (and remains) incisive, and located the primary challenge as a need for new kinds of programming tools that would go beyond the visual bias of computer graphics software. VR worlds, Leler pointed out, are assumed generally to be made first using computer-aided design (CAD) software, which "encourages us to think of the virtual environment as an entity that exists independently of the observer (like a virtual building to be walked through)."

> The resulting environments are largely static with minimal user input; any interaction is typically bolted on after the environment is "created" by the designer. Even what is considered advanced interaction, such as hitting virtual drums to make sounds, is derived from collision and interference detection algorithms developed for CAD use. More typically, the user's role is limited to putting on the eyephones and exploring; after an environment has been explored a few times it often becomes boring.[30]

Leler went on to make an extended comparison between this oculocentric bias and advanced work happening in electronic music, some of which he observed firsthand in the Banff Centre's media arts program. Referring to interactive music systems made by Daniel Scheidt and George Lewis, Leler observed that "in some sense these are [also] virtual environments: the computer program takes input data from the musician, interprets this data in some way, and then generates a response, in the form of audible output. This [output] in turn, is heard by the musician and affects what is played. The role of the composer/programmer is to define how the musical gestures of the musician are interpreted and the nature of the (musical) response to these stimuli."[31]

Treating each of these mediations as comparable to a filter, Leler then argues that both can possess "behavior," and thus can be considered as "actors." Leler's conception of the elements of an environment as distributed, interdependent actors imbued with behavior had captivated the Media Arts team, to the extent that his diagrams and design philosophy served as the technical foundation of the project proposal to the federal government.[32] Yet Leler's vision of the VR user acting also as a designer capable of changing the virtual environment while occupying it did not become the basis for subsequent system development.[33] Thirty years later, programming VR "from within," at least in the musically enlivened way envisaged by Leler, is still a developing research frontier.[34]

Inherent Rights, Vision Rights (1992)

During the Bioapparatus residency, half a dozen VR artworks were produced using the basic system that had been obtained within the limited resources of the startup grant. Still the system permitted full immersion using a head-mounted eyephone display and spatialized sound, albeit limited to update rates of less than ten frames per second. Lawrence Paul Yuxweluptun, the only indigenous Canadian in the project, called VR a "white man's mask . . . employing technology that has been used against native people . . . to show people what is happening to me spiritually."[35] Yuxweluptun began his exploratory VR work during the height of a hot period of political conflict between native groups across Canada and the federal government, arising out of a land dispute outside Montréal between Mohawk people and the neighboring town. His starting point was to critique the VR format, not just as a specific representational template, as Richards and Tenhaaf had done, but in broader terms as a symptom of the colonial repression of indigenous peoples. As a painter, Yuxweluptun understood his depictions of environmental degradation and ecological despoilment, of the "schumck" brought to his people by colonial occupation, as a form of political activism. Similarly, while also viewing VR as the "experience of a new art form," he found it could be redirected within his artistic construct in a way that was continuous with his assertion of art making as an "inherent right"—in this case a right to continue his visionary experiences as a native participant in spiritual quests. This spiritual dimension was presented in his VR piece as a "possessed state of feeling rhythmic sounds in a longhouse, feeling sounds go through one's own self, feeling a spirit inside you." Compensating for the visual crudeness of the low-end VR system through high sonic fidelity, Yuxweluptun generously invited his audience "into contact with a native worshiping way of life, praying Indians—a way to bring others close to my heart so they can understand my belief system."[36]

When it came to exhibit the work in galleries, the Banff team built a static viewfinder to replace the bulky, uncomfortable head-mounted eyephones, with a hand controller for navigation. The work was first shown at the National Gallery of Canada in Ottawa, as part of a landmark 1992 exhibition on First Nations art, *Land Spirit Power*. As understood by Diana Nemiroff, chief curator of Canada's most prestigious art museum, the work of Yuxweluptun was capable of moving outside the media arts framework, to occupy a discursive space concerned with "the issue of colonialism," the loss of "a sense of the sacred," and the "ongoing struggle of the aboriginal people of North America for their ancestral lands."[37] *Inherent Rights, Vision Rights* was displayed again in 2016 as part of a solo retrospective exhibition at the Museum of Anthropology in Vancouver.[38]

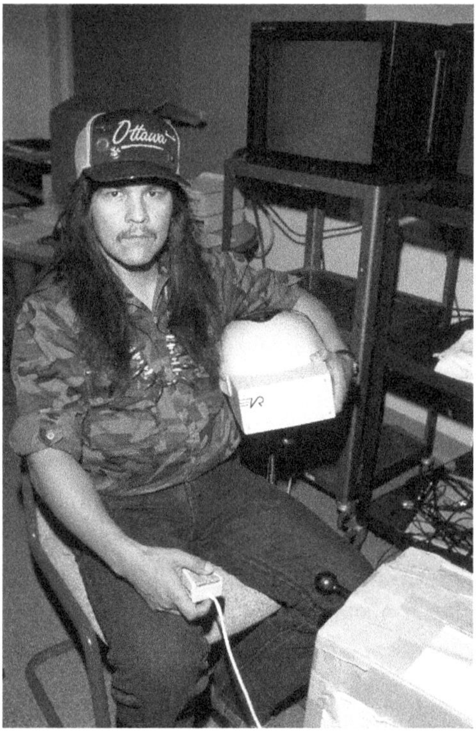

Figure 7.1
Lawrence Paul Yuxweluptun at Banff, 1992.

Spectral Bodies (1991)

Catherine Richards's mock-scientific speculative-fiction video, *Spectral Bodies*, was shot in Edmonton at the laboratory of the Computer Science Department of the University of Alberta; postproduction occurred in the Banff Media Arts facilities. In the artist's words, the piece "considers the simulation and subjectivity in VR technology through narratives about inhabiting our bodies (proprioception) and losing this sense."[39] It is therefore consistent with the research questions about body illusions that motivated her first interest in VR as a template for interrogating what it means to for the body to be in close symbiosis with the machine.[40] At the center of the piece is the image of a spectral hand and arm, depicted in low resolution as dots and skeletal lines. These dots and skeletal lines are moved by a VR data glove. The video depicts the enactment of physiological experiments, with the aim of "creating body illusions with 'subjects' [to] suggest that our imaginary bodies (or are they real bodies?) can be destabilized."[41]

The work has been insightfully analyzed by Australian media philosopher Frances Dyson. For Dyson, Richards treats VR as a philosophical instrument for both "creating and interpreting narratives about proprioception, not as a medical or scientific actuality, but as a fiction that forms narratives, which themselves spread across personal histories and scientific research."[42] In this context, VR acts as an instrument for "questioning the tight intertwining between our instruments, realities, sense(s), and fictions of self." Importantly, she underlines Richards's own assertion that after they are destabilized in the ways that are depicted experimentally, they are ready to be remapped figuratively, showing that "the boundaries of the body are permeable, that the unified autonomous subject is a myth."[43] *Spectral Bodies* is also a remarkably prescient speculative artwork, when seen nearly thirty years later against contemporary virtual embodiment research in psychology and philosophy.[44]

Art and Virtual Environments, Redux

Douglas MacLeod, the managing producer of the nine second-stage Art and Virtual Environments productions, recollects how complex the project was to direct: "Chaos reigned. We really had no idea what we were doing. . . . It was like staging nine different operas in two years while at the same time trying to invent the idea of opera."[45] MacLeod was one of the contributing editors of the 1996 MIT Press volume that offers the artists' statements, their production credits, and critical essays by notable theorists and scholars, including N. Katherine Hayles, Cameron Bailey, Frances Dyson, Erkki Huhtamo, Margaret Morse, Loretta Todd, Jeanne Randolph, and Avital Ronell. The Banff-produced works were also given further consideration in scattered publications over the next few years.[46] However, only occasionally have these works been referred to during the second springtime of VR, which kicked into gear seriously in 2014 with Facebook's multibillion dollar investment in Oculus Rift, an eyephone equipment maker. Compendia and textbooks aiming at full coverage of digital and so-called "new media" art since the millennium tend only to register a single contribution from the first springtime of VR analyzed in this chapter, namely Char Davies's pair of VR installations, *Osmose* and *Éphemère* (1995, 1998).

While fully deserving this scholarly attention, which has focused principally on Davies's use of a breath-based interface, kinesthetics and balance for navigation, and her replacement of hard-edge surfaces with layered translucency, little recognition has been given so far to the crucial role of the production team that Davies hired to support their development. Davies's team at the time included John Harrison, Glen Fraser, Dorota Blaszczak, and Rick Bidlack; all had worked on the Banff Centre's VR productions, with the first two providing the VR programming and the second pair the sound design and engineering. Davies had initially intended to attend the Bioapparatus residency and did not only because of illness. In her letter of inquiry to studio art head Lorne Falk, Davies aligned her interest directly with the aims of the residency, i.e., "to take a wary and critical stance toward the technology, particularly in relation to the body." *Osmose*, still the most widely discussed, indeed legendary early VR artwork (though it has not been shown publicly for the past 15 years), is in this sense both continuous with the contested concerns of the Bioapparatus residency, while also growing directly out of the technical and creative skillsets developed in the second, large-scale production phase.

Only one participant in the Bioapparatus was also commissioned to produce a work in the second phase, and it is not surprising that this was the individual who self-identifies most explicitly as a personality straddling the worlds of art and research. Michael Naimark also had little patience for the state of the VR technology itself, though he made ample use of the Banff Centre's Silicon Graphics ONYX Reality Engines, the fastest graphics computers generally available at the time. A graduate of the Center for Advanced Visual Studies, the Massachusetts Institute of Technology's graduate program in technological art, Naimark also was a contributor to the visualization research carried out in the Architecture Machine Group, much of it funded by the Pentagon's Advanced Research Projects Agency. As a filmmaker-cinematographer, with an abiding interest in the immersive documentation of "real space," Naimark accordingly applied some of the lessons from the famous Aspen Movie Map to create several field studies at Banff, in the process devising techniques for combining three-dimensional computer models with images of the physical world. Criticizing the default in the computer world for building models from scratch rather than from cameras, and the resulting bias for "fantasy worlds, imaginary places, and scientific visualizations," Naimark's efforts were divided between the field recording studies as such, and the creation of a "media archeological" *détournement* of virtual reality itself.[47] *See Banff!* is an interactive stereoscopic Kinetoscope installation operated with a crank, allowing the viewer to scroll back and forward along preset paths offering views of the Alberta landscape. Naimark was then employed as a staff researcher at Interval Research Corporation, a Palo Alto firm headed by Microsoft cofounder Paul Allen and dedicated to looking forward five to ten years into the future of computing and media. Naimark's project overlapped with Interval's interests in computer vision and other scientific matters, but he segregated *See Banff!* successfully as an independent art project.

The Interval corporation also played an intensive role in the largest of the Banff VR efforts, Rachel Strickland and Brenda Laurel's epic experiment in "narrative play," *Placeholder*. Both women were on Interval's staff, as was engineer Rob Tow; ultimately, the work was co-produced by the corporation and the Banff Centre. Laurel, a renowned human-interface design expert, and Strickland, a documentary filmmaker by training, were the primary creative authors of the *Placeholder* scenario, which placed spirit characters into the virtual world so that people could enter their bodies, thus to enable them to experience flight as a crow, or see the world as a snake, and so forth. The "smart costumes" invented for this work were set into an evocative spatial audio environment, which provided the ability for users to leave vocal traces at chosen points in virtual space so that future visitors could listen to the accumulating memories of the space. The developers ran out of time to implement the voice-holder feature, an indication of how far the ambition of its creators outran the available time and resources. *Placeholder* was also controversial for allegedly appropriating indigenous stories, and while Laurel and Strickland tried to adapt their use of mythology in a culturally sensitive way, the tension arising from these charges of appropriation flared up again at the closing symposium where the works were collectively presented.[48]

For some, the participation of a prominent Silicon Valley research corporation in Banff's VR project was controversial for another reason. While some saw it as a clear indicator of the relevance of the project overall to the wider world of computational media research, others pointed out that Banff's initiative could be construed more accurately as a "cultural futures market of governmental and industrial sponsorship" than as a field of creative experimentation. The author of these words was Ron Kuivila, a composer and sound artist, who went on to note that "the ideology underlying cultural sponsorship that intends to promote industrial growth warrants careful scrutiny."[49]

Kuivila's *Virtual Reality on $5 a Day* does just this: it "attempts to critique VR, the movie, while exploring VR, the Gizmo."[50] Kuivila assailed the installation's visitors with a raucous flock of virtual agents flying around their heads, singing, jabbering and barking, and feeding on the virtual detritus left by visitors' motions. The piece is interrupted by advertising messages from the fictive industrial sponsor of the project, IO (Informatique Oblique). In a chilling anticipation of the coming gamification of education and business, one of IO's advertising pitches is cheerily announced in this way:

> Chastened by the failure of artificial intelligence, encouraged by our faith in human intelligence, we at Information Oblique . . . showed that game space is a dual space for the *problem set of humanity*. [At IO] . . . the new science of *Ludic Analysis* allows any problem or task to be re-expressed as an interactive game. Once ludified, the problem is posted to our Intelligence Mine, where millions of crack game players from our Intelligence Miners, can have a go at it.[51]

Also taking a parodic stance, though one less cynical than Kuivila's, multimedia artist Perry Hoberman created an interactive installation for multiple participants by papering an entire gallery space with printed bar code symbols, making every surface responsive. Thus, the interface itself became immersive, with the triggered codes generating two kinds of response. First, new objects are "born," appearing as cartoonish three-dimensional models of everyday objects—hats, shoes, or sunglasses—while other codes are scanned to incite movement instructions to these objects, making them expand, contract, breathe, tremble, jitter or bounce. Hoberman's work was of immediate public appeal and one of the few Banff VR projects that received much exposure in exhibitions elsewhere; for critic Jeanne Randolph, the piece delighted as would an "exuberant, misbehaving pet."[52] According to Randolph, "delusions of ecstasy, freedom and omniscience with which virtual reality has been promoted were conspicuous by their absence."[53] Artist Rafael Lozano-Hemmer placed Hoberman's work near the top of his list of media art productions that successfully "pervert technological correctness": the "misemployment of barcode technology, which was originally designed to read marketing information, gives way to stimulating interaction with virtual objects reminiscent of René Magritte's aesthetics."[54]

Montréal playwright Michael MacKenzie (one of the few Canadians included as a participant in this full production phase) teamed with New Yorker Toni Dove to create a complex VR combining interactive graphics and a commissioned sound score within a multimedia theater scenario. *Archeology of a Mother Tongue* was the only one of the Banff VR projects that aimed to create the feeling of being inside an authored narrative space, of taking the part of a character walking around in a movie. The scenario sets the viewer initially within a dream, then places her as coroner in a murder investigation unfolding in a gritty, multilayered landscape mixing together wire-frame models of imaginary architecture by Piranesi and grainy video material piped into the VR world from optical disks. On the occasion of a 2018 remounting of this installation, composer and author George Lewis commented on the work as a sophisticated experiment in narrative form: "Navigation among prefabricated pathways, or what Dove called 'the fallacy of choice,' was replaced by an . . . immersive flow that provoked a more fluid sense of time."[55]

A deep concern with liquid time, though developed in a more specifically musical sense, permeates the VR work produced by architect and sound artist Marcos Novak. Novak had already theorized "cyberspace" as a form of liquid architecture,[56] and for his efforts at Banff, which made use of the canonical immersive head-mounted helmet, Novak offered another neologism: "ArchiMusic."[57] For Novak, this term designates a kind of navigable music that is heard while moving across a "slowly evolving dataspace, composed as a multidimensional surface. . . . What is heard at any given time is the outcome of the viewer's trajectory. The compositional algorithms track the viewer, also triggering visual events in the form of semiautonomous dancing forms."[58] Providing both the software code and design, Novak was assisted in this foray into sonic virtual-world making by composer Rick Bidlack, who would join Char Davies at Softimage in Montréal the next year.

Connecting directly with the interrogation of virtual embodiment in the seminar on the Bioapparatus, designer Diana Gromala and choreographer Yakov Sharir created a multicomponent work presented both as a multimedia dance performance and a single-user VR experience. Viewers navigated virtually through the interior of Gromala's torso, gathered from MRI data, capturing fleeting glimpses of poetic texts reflecting on her experience of chronic pain, and also of Sharir's whirling body, intended to be suggestive, as indicated in the work's title, of "dancing with the virtual dervish." As noted by media theorist Mark Hansen, Gromala's and Sharir's meditations on the virtual body "position the aesthetic potential of VR in stark contrast to the rhetoric of disembodiment and transcendence so frequently associated with it.[59]" Gromala went on to develop a continuing research program on the "transformation of pain," combining immersive VR with biofeedback techniques.[60]

When Banff Centre management was asked after the end of the Art and Virtual Environments Project to account for the fact that a small minority of the participating artists were of Canadian nationality, the Centre responded with a holistic account of the benefits of the project.[61] The artists were not the only beneficiaries of the investment of resources, Vice President Carol Phillips wrote, but also the mainly Canadian technical teams, including programmers and engineers who had been called on to figure out how to solve the problems of actually executing the varied visions of the artists. If, in MacLeod's image, producing the Art and Virtual Environments artworks was really like inventing "the idea of opera," the fundamental stage magic was provided by this hardy group of software developers and engineers.

When Char Davies hired the Banff Centre's lead VR software architect and analyst, John Harrison, to be the first member of the team producing *Osmose,* she was employed at Softimage as director of visual research. As seen in chapter 3, Daniel Langlois had founded the firm in 1986 with substantial support from Davies in the form of organizational and user-interface design as well as business expertise.[62] Davies's abiding interest lay in extending the imaginative scope of 3D computer graphics, as she wrote in 1991 to Lorne Falk, to "compensate for the technology's ideological tendency . . . to lead us away from bodily experience of the world. By subverting technology's conventions (photo-realism, linear perspective, Cartesian space, etc), I am able to use it to emphasize a corporeal, organic aspect of reality in which mind and body—subject and world—are merged."[63] Harrison, in his turn, had learned deeply from the challenges of the Banff VR program; yet, by the time of the program's conclusion, he was ready to focus deeply on a single project formulated explicitly around Davies's well-researched aesthetic: not to reproduce the world, but to layer it with transparency and diffuse light; not to point and select, but to glide using breath and balance.[64]

As already indicated, the writings about Davies's VR achievements are substantial, covering in detail its interactive features and theories of embodiment,[65] its place within the art history of virtual spaces,[66] and its pertinence to a sonic philosophy of media,[67] to mention only a few of the scholarly treatments. By 1994, software giant Microsoft had acquired Softimage, and the prominent corporate sponsorship of *Osmose* came to the fore when the work was premiered at the Museum of Contemporary Art in Montréal, a showcase exhibition within the 1995 International Symposium on Electronic Art. In stark contrast to the overwhelmingly positive reviews *Osmose* garnered there, which sometimes bordered on hushed reverence, a strong dissent was penned by an Australian techno-art group called the Bureau of Inverse Technology.[68] Setting aside its vitriol and *ad hominem* attacks, the BIT ambush of *Osmose* rings true today in one sense that has nothing to do with the distaste its authors felt for the work's philosophical aspirations, or for the spiritually redemptive testimonials voiced by viewers. In 1995, VR art at the level of production values Char Davies had access to—or the Banff Centre had earlier in the decade—was inaccessible to anyone without immense resources and technical expertise. In those years, not only was the medium in its infancy, but huge sums were needed to experiment with it at all. Twenty-five years later, the production capabilities are within reach of any reasonably equipped independent studio. The aesthetic and philosophical questions around VR, enjoined with vigor at Banff and carried forward by Davies in the 1990s, continue to animate her current research on the virtual reality of natural forest worlds around her Québec home.[69]

As one of its instigators, I look back on the Banff Centre's multiphase, critically inflected project on art and virtual environments with a mixture of impressions. Emphasizing embodied interaction, multisensorial interactivity, and the fluidity of body-machine boundaries, the productions taken as a whole strike a number of common chords with the cases treated in the previous chapters of this book. Yet the impact of most of them was muted: the hardware and software needed to run them remained widely inaccessible for decades, while the very instability of their setups consigned most to the status of performance demos.[70] VR lost its allure as cybercultural mythos, while Canadian policy priorities turned to information superhighways. The computer-generated virtual worlds explored at the Banff Centre would become networked metaverses by the 2010s, advanced crucially by the computer game industry and growing applications in health care, engineering, and education,[71] with growing acceptance in the fine art world as well.[72] The Banff Centre's effort to align an anticipatory critique of a new technological imaginary with the innovative interests of government funders, academic researchers, and business firms registers now as indicative of the difficulty of harnessing such a disparate, potentially antagonistic set of energies. The next chapter recasts the complexity of this challenge more broadly against the disjunctive forces of globalizing information infrastructures and shifting domestic cultural politics.

8 Après le Déluge: Periodizing Canada's Episode of Light

In the opening chapter of this book, I represented the period from the mid-1960s to early 1990s with the stylized image of a pair of overlapping arcs—one rising, one falling. The first of these arcs tracks the origins and early deployment of new media and the digital arts in Canada, propelled by the singular surge of national élan following Expo 67 and ascending to the mass diffusion phase of the "digital revolution" in the 1990s. The second, descending arc follows the evolving national politics in Canada, likewise passing through the heady enthusiasms of Expo 67, but this time tracking a declining figure of disillusionment and existential crisis that reaches its nadir in the 1990s, when Canada faced severe doubts about its continued survival. While the rising arc is global in scope, the declining arc is unique to Canada. The divergence captured in this image defines the end of the period covered in this book—when the digital revolution explodes into the mainstream, just as Canadian identity appears diminished, fractured, and uncertain—and prefigures the opportunities and challenges that would face Canada's creative and research community in the following decades.

I have characterized what was most distinctive about the Canadian transition into the era of digital media in terms of an "alternative technological ethos," a concept developed across this book in two complementary ways. First, as a characteristic spirit of experimental discovery-in-use, it designates a method for approaching new technological means that favors sensorial immediacy and improvisatory expressiveness, prior to the closure and standardization that comes with commodification and mass adoption. In this sense, it contrasts with the kind of normative, instrumental consideration of the potential of a new technology, which seeks solutions to predetermined problems and where a technology's value is judged on the basis of optimization metrics. Second, this alternative technological ethos has appeared in the preceding chapters as a component in the wider set of factors shaping multifaceted innovative outcomes, unconfined by field, form, or practice, spanning out to foster the coalescence of new interdisciplinary fields, research paradigms, and models for digital sociality. Norman McLaren's instrumental, hand-made cinema, for instance, left traces of an innovative proto-computational approach to filmmaking, demonstrating a digital sensibility *avant la lettre*, and in the process helped to actualize research in computer graphics through its impact on the computational imaginary of the emerging community of software engineers. In a similar vein, the previous chapter revealed how virtual reality in Canada was framed and contested from philosophical and critical as much as engineering and artistic perspectives, through a common, underlying creative method emphasizing embodied interaction and multisensorial expression. In this concluding chapter, I want to draw together these parallel strands of the alternative technological ethos—as both an exploratory method and a model of interdisciplinarity—in order to clarify the wider significance of the story of Canadian innovation during the years of transition into the digital era that marks the conclusion of the period explored in the preceding chapters.

At the media-technological level, the distinctively Canadian alternative ethos that I have described was initially fostered by its appearance at a moment, in the late 1960s, defined by an expansive unbinding of national-political energies in Canada. This supportive environment for innovative media-artistic initiatives was never deliberately realized as a consciously designed policy framework; rather, it emerged out of a decentralized constellation of temporary, often ad hoc projects and programs over the 1970s and 1980s. In the 1990s, as information technology, repackaged with a neoliberal ideology of creative industries, swept the planet, "new media innovation" itself became a cognizable policy objective. While media convergence and ubiquitous connectivity had been anticipated since the 1970s in Canadian government reports as well as academic and popular writing on culture and communication, official attempts to formulate innovation policy, upon the full arrival of the internet and media convergence on a mass scale during the 1990s, were handicapped by bureaucratic miscues and confusion. In this contentious context, intensified by a fraught political climate, the emerging community of new media and arts practice sought out and often found novel resources and strategic policy spaces to support their work. This entailed realigning creative arts practice within the traditional research institutions, like granting councils and university-based graduate programs, finessing creative-economy incentives, and mainstreaming digital media in the wider field of the established arts. The full details of these efforts at policy and disciplinary realignment lie beyond the scope of this chapter, other than to briefly note their thematic continuities with the alternative technological ethos that characterized the preceding period described in this book.

The first of these continuities, apparent across the cases explored in this book, lies in the way creative agency is distributed between people and technologies. Canadian prominence in human-computer interaction (chapters 3 and 4), as well Canadians' key contribution to the origins of acoustic ecology (chapter 5), arose from a nation that placed a special emphasis on communication in both the quotidian and the practical, nation-building senses. At the fine scale of creative practice, Canadians evinced an early grasp of software as an expressive medium and of user interfaces as playable thresholds. More broadly, in 1993, during the high tensions of its disunity crisis, B. W. Powe went further to call Canada a "communication state" *tout court*, "an experiment in alternate current . . . [in a] condition of receptivity . . . mediated through technology."[1] Urging Canadians in that fraught period to consider their country less as a nation-state and more as a "state in process," Powe wrote that rather than insisting on a monolithic conception of political unity, Canadians should "welcome what is multifarious, flexible, personalized, paradoxical and protean" in their national character and accomplishments.[2] This processual idea of Canadian identity as plural and polyphonic can be read now as a valedictory, even elegiac statement of loss for what was felt to be a threatened set of supposedly national traits. Yet a penchant for finding meaning in the "resonant interval," the gaps *between* rather than within the essential natures of people, things, ideas, or institutions, was constitutive of the McLuhan-inflected creative atmosphere of the 1970s and 1980s in Canada. The alternative technological ethos, with its focus on enriching media interface and interaction as a "fertile nexus" for creative practice, registers throughout this book as homologous to this processual, experimental idea of Canada as a communication state.

A second continuity locates an emphasis on sensorial immediacy and embodied interaction within an organizational frame encouraging the interpenetration of art and research. The worlds of art, engineering, and science came together around research objects imbued with a certain "interpretive flexibility" that allowed groups to work together without necessarily sharing an explicit consensus.[3] An early fostering of computer music, animation, and dance as driving elements of computer science research programs, on one hand, and the experimental development of networked, interactive and immersive environments within art studios and cultural institutions, on the other, are scattered across the chapters in this book. In the standout case of a squandered opportunity, the absence of involvement by creative users in the design phase of the Telidon project stands apart from a general tendency favoring mutually informing interconnections between art, science, design, and engineering. As I have already noted, the organizational context in which this propensity toward artistic research occurred was typically ad hoc, and sometimes not even well understood by its instigators: early breakthroughs in creative applications for interactive computing at the NRC, for instance, were not sustained after what was understood simply to be an interesting proof of concept.

Both of these thematic continuities, that of distributed creative agency and the co-production of art and research, appear across the full range of art forms. The figure of the artist as creator of "counter-environments,"[4] a term McLuhan also applied to Canada as a whole, is a third thematic continuity linking the period described in this book to the years that follow. Artworks can reveal and perturb the invisible effects of media environments. Already at the dawn of the digital era, the ecological approach to sound and music fostered by the World Soundscape Project (chapter 5), Gould's emphasis on reconfigured "structures of participation" afforded through new music technologies (chapter 5), McLaren's counter-Hollywood paradigm of artisanal animation (chapter 2), and Canadian artists' experiments with telecommunications as a performative milieu (chapter 6) all ran counter to the mainstream cultural or corporate environments within and against which they were posed; yet they were *also* prefigurative of the rising computational sensibility to come. By the 1990s, the arts would be more deliberately incorporated in research and innovation policy, though in the Canadian instance, not without considerable confusion.

Creative Industry Becomes Global Orthodoxy

For historian Eric Hobsbawm, the "short twentieth century" ended in 1989, demarcated by the end of the Cold War and the attendant triumph of neoliberalism, the latter vitally propelled by the Information Paradigm itself.[5] The experimental, exploratory cultural uses of new media, seen here through a primarily Canadian lens, grew from a rising wave into a global deluge. In the jargon of the 1990s, "content" was king. Convergence—in various guises, generally entailing the merging of the content industries with the tech sector—was driving the wildest of economic get-rich fantasies. Executives with the slightest connection to the arts were positioning themselves alongside the "creatives" who would, ostensibly, become key components in their new-economy strategies. Art came to be understood as an instrumental input to the content pipeline for new technological platforms: multimedia, networked, interactive, immersive. A flavor of this consideration of the arts from an engineering standpoint can be found in Patrick Purcell's 1997 argument in favor of "the primacy of content in digital media."[6]

> The success of this content-led approach . . . will depend on a close working relationship between content creation and technology in the form of new platforms and advanced communication systems. It calls for the content creation community to take a more robust and assertive stance—"pole position"—in research. The technology disciplines must therefore jettison outmoded professional cultural baggage and foster a new cadre of content makers with radical ideas.[7]

Television, for instance, could be reenvisioned in ways that go beyond higher-definition displays or more program selections, as Purcell observed that the "incipient feedback" then available could lead to "innovative forms of content."[8] It was during these same years, while art was being formulated as a research and development input, that the policy discourse of "creative industries," inflected by and serving the ends of the Information Paradigm, was coming to the fore, first in Britain and then around the world.

The Blair government's initial formulation of the concept of creative industries in 1997 was as follows: "Those activities which have their origin in individual creativity, skill and talent and which have a potential for wealth and job creation through the generation and exploitation of intellectual property."[9] Explicitly, "creative industries" embraced the arts and media, museums, archives, and the traditional mass cultural industries, but just as crucially software, design, and advertising. The phrase spread quickly, with sober economic treatments, like Richard Caves's eponymous 2000 volume that emphasized "contracts between art and commerce," and other breathless celebrations of the "creative class" as the panacea for all economic and social ills.[10] In 2003, the US National Academy of Sciences sought to align information technology with the arts and media, dubbing them all an "exciting new domain" of "creative practices" that went beyond the supposed productivity gains usually associated with, and expected of, computerization.[11] Stuart Cunningham, a key framer of creative industries policy discourses, wrote in 2005 that "creative disciplines" can no longer be relegated just to the realm of "civilizing" activity.

> Instead they must be recognized as one of the vanguards of the new economy. R&D strategies must work to *catch the emerging wave of innovation needed to meet demand for content* creation in entertainment, education, and health information, and to build and exploit universal broadband architectures in strategic partnerships with industry.[12]

Embraced by the European Union, China, and the United Nations, by 2015 creative industry policy had become, in the phrase of Philip Schlesinger, a "global orthodoxy."[13] For critics of this orthodoxy, the absorption of the arts into the creative economy reduced the assumed intrinsic value of culture to its economic contribution, which is hardly a new story; indeed, in the Canadian case, this reframing of the arts and media occurred at a particularly fraught moment.

Just as Canada crossed the threshold out of Hobsbawm's truncated twentieth century, it was experiencing a plummeting sense of itself as a sustainable national entity. Writing at the time from the standpoint of his own discipline of history, Michael Bliss diagnosed a turn away from issues of broad national scope, toward matters of a more private scale, like ethnicity, family, class, and gender. "The sundering of a sense of Canadian history," he argued in 1992, "became part and parcel of the sundering of Canadians' consciousness of themselves as a people."[14] The very years that had elsewhere been welcomed triumphally as the end of history were, in Canada, suffused with existential angst, often expressed in terms of a search to define an essential, and therefore presumably unifying, concept of Canadian identity.

This was most clearly articulated though the obvious contrasts with the United States, against which Canada could still appear to be a more tolerant society, less conflictual and violence-ridden, and more committed to collective provision and diversity. For philosopher Charles Taylor, writing in 1992, accommodating Québec's aspirations as a nation, alongside those of indigenous peoples, would require going beyond this elementary level of tolerance for diversity, to what he called "deep diversity," where "a plurality of ways of belonging would also be acknowledged and accepted."[15] Yet at the time, Taylor was pessimistic, harboring fears that the country was actually "drifting towards break-up," with "little being done to prepare for this, let alone head it off."[16]

Not surprisingly, seen against this broader canvas, the clarity of Canadian policy aims during these years was undermined by an unstable tension between measures conceived to keep the country together and a pragmatic embrace of the established cultural industries as competitive exporter in the emerging neoliberal international order.[17] During the negotiations beginning in the 1980s leading eventually to the 1994 North American Free Trade Agreement with the United States and Mexico, Canada fought for and won an exemption to permit continued direct and indirect subsidy of its cultural industries. Remarking on this maintenance of the cultural exemption, Alison Beale observed that for Canadians, cultural sovereignty is tantamount to national security.[18] This hard-won concession was primarily in support of the old media that comprise the usual grouping of cultural industries—broadcasting, cinema, publishing, and sound recording—producers of what McLuhan dismissed as "packaged programming, nineteenth-century style":[19] commodified and transmissible in the mass media. The irony of this emphasis on safeguarding old media should now be evident in light of the abundance of examples presented in this book that feature an exceptionally early Canadian competence with the affordances of the emerging new media: interactivity, participation, connectivity, and immersion.

In many ways the instrumentalization of the arts and media that is constitutive of creative-economy discourse and policy was, by the 1990s, already long established in Canada. As I described in chapter 1, by the 1990s the mandates of Canada's central cultural agencies and its media and arts funding policies had become common knowledge to Canadians, who at least from the 1930s looked favorably upon publicly supported provision of culture as a bulwark against the deluge of American commercial programming. The particular conception of identity-reinforcing culture that came to the fore in the 1990s, however, appears to have been accentuated by the dissolution in 1993 of the federal Department of Communications, which reassigned responsibility for arts and culture to a new Department of Canadian Heritage, and technology and research mandates to the Department of Industry. In mandating the "promotion, dissemination and preservation of stories and symbols reflective of the past and expressive of our values and aspirations," Canadian Heritage adopted a strikingly retrogressive stance, particularly in light of the Department of Communications' earlier efforts, however ineffective they sometimes were, to unify technological and cultural concerns "under one roof."[20] A high-profile federal commission established in 1994 to study and make recommendations on the "information highway" is a good example. This commission struck the defensive posture of traditional cultural policy discourses when it adopted as one of its pillars the reinforcement of Canadian sovereignty and cultural identity, alongside universal access, job creation, and investment in the information technology sector. This turning backward to identitarian notions of culture at just the time when a more synergistic approach to content-led research might have been prioritized reflects the relative dislocation of Canadian policy thinking, a sense in which time was out of joint during the last decade of the century. Within this anachronistic policy context, where heritage concepts of Canadian identity were emphasized alongside a focus on old-media cultural exports, Canadian success in the animation software and production industries could be duly celebrated, heralded by the foundation in the 1980s of firms like Softimage, Alias, Discrete Logic, and SideEffects, though most of these had already been sold to American multinational corporations in the 1990s.

Against this anachronistic backdrop, efforts were made to define new media research itself as a policy object—an effort I contributed to while serving as a consultant to Industry Canada and Canadian Heritage. These proposals included the approval of public support for hybrid research networks, which aimed to group together university-based research with the independent media arts scene and artist-run center networks, alongside industry partners, national labs, and cultural agencies. Other proposals advocated the placement of artists in science labs, and new funding programs more generally seeking to reposition the creative arts explicitly as research so as to make them more fundable by non-arts funding bodies. The success of these federal policy initiatives was varied. Between 2002 and 2005 regional research networks for new media were supported directly by Canadian Heritage. From the federal flagship program for research centers of excellence, a network for computer graphics and animation was funded from 2009 to 2015.[21] In 2002, the Social Science and Humanities Research Council launched regular granting programs for "research-creation," following the example previously set by the government of Québec, with the aim of legitimizing artistic practice as a kind of academically acceptable form of knowledge production.

Hexagram, an interuniversity network for creation in media arts and technologies launched in 2002, was an early example of such "research-creation" initiatives. Founded in 2002 through a partnership between Concordia University and the University of Québec, Hexagram began with a broad mission to establish close relations between the worlds of academia, business, and the technological arts. Indicative of the mixed effectiveness of these kinds of linkages more broadly, studies of Hexagram's first decade of operations pointed out the dilemmas artists faced when confronted with the sometimes-contradictory imperatives of Hexagram's "double rationality, at once artistic and entrepreneurial."[22] For sociologist Jean-Paul Fourmentraux, the duality he found inside Hexagram represented a "primordial challenge," when seen in the context of a justifying discourse that is at once a "gamble and promise," combining "rational strategic forecasting (development of markets, technology, applications) with speculative conjecture that belongs more in the realm of social utopia and literary and artistic imagination."[23] In another case of bridging between art, research, and industry, the Banff Centre widened its focus beyond VR to embrace the burgeoning digital content industries, first in an effort to develop multimedia authoring tools for artists,[24] and then in its Institute for New Media, founded in 1996, which for a lively decade supported high-powered summit meetings and catalyzed collaborations involving researchers, artists, business firms, and policymakers.[25]

It will be noted that many of the research initiatives just mentioned were launched *after* the dot-com crash of 2001. Canada had been insulated somewhat from this recession, having engaged already in the mid-1990s in a deep austerity program that, by the 2000s, allowed for budget surpluses to be channeled into renewed cultural and research funding. As in previous moments discussed in this book, the 2000s saw comparatively generous state support given to experimental media arts production, in the form of direct subsidies to individuals, R&D tax credits, and jobs programs, along with the academic institutionalization of artistic research.

An understanding of artistic research as a distinctive form of knowledge production continued to develop along several trajectories in the 2000s. Under the rubric of "research-creation" many Canadian universities have sought to leverage digital media practice as knowledge production and to reclassify graduate education in creative arts fields as the training of "highly qualified personnel." In its implementation at the federal level of Canada's Social Science and Humanities granting council, the criteria for research-creation were replete with accountability-grounded requirements for a well-defined "scholarly apparatus," "clear research problems," and a rigorous "methodological approach" to the artistic creation process.[26] Research-creation when formalized in this way provides crucial support to sustain long-range, blended academic-artistic inquiry in emerging fields, while, for less sympathetic critics, it risks being milked as a haven providing "props to bad artists."[27]

Henk Borgdorff has described artistic research within a European academic context as a "conflict of the faculties," urging a consideration of the intrinsic value of artworks themselves as "non-discursive knowledge" that challenges traditional disciplinary monopolies and epistemologies.[28] Less abstractly, Concordia University media artist and scholar Chris Salter and coauthors coming from science and technology studies offer a survey of topics and sites for critically grounded artistic knowledge production, such as performativity and improvisation, sensorial engagement, social intervention, and "emergent biosensibilities." As Salter notes, this last "heterogeneous category of artistic work with living systems . . . is carried out in both science labs and art studios, taught in art schools, and increasingly institutionalized in special programs and academic centers."[29] These biosensitive concerns with living systems can be understood as carrying forward the alternative technological ethos that figures so prominently in the mid-to-late-twentieth-century transition to a computational sensibility documented in this book.

This book has provided a detailed history of the way Canadian artists catalyzed innovation during the early, developmental phase of a new techno-economic paradigm. From a more distant vantage, this was a period of "decompartmentalization," a term used by art historian Erwin Panofsky to refer to the European Renaissance, when new "transmission belts" between domains of knowledge and between artists, scientists, and engineers enabled novel "methods of picturing and comprehending the world."[30] The conjuncture in Canada of this kind of techno-aesthetic fluidity with a short-lived burst of political self-reinvention was followed by a complex overlay of forces. The country adapted to globalizing pressures while the digital became ubiquitous and artistic research itself was institutionalized academically. Today, advocates for an explicitly "postdigital" agenda for art and technology turn away from what are seen as the depleted attraction of dematerialized virtuality, life on the screen, and surveillance capitalism, urging instead the deconstruction of distinctions between digital and analog, old and new media, and the celebration of "DIY agency" outside "totalitarian innovation."[31] This call for a different set of research values in the arts, media, and technology is occurring just as voices in the cultural policy research community like Philip Schlesinger are calling for countervailing values to oppose the reductionism and simplifications of the creative-economy orthodoxy. Drawing on a 2016 British government white paper, Schlesinger notes a renewed recognition of the "intrinsic, enriching value of culture in and of itself," and "the social value of improving educational attainment and health."[32] Yet he allows, with a whiff of exasperation, that returning to older logics of intrinsic artistic elevation or social utility does not, on its own, go very far toward generating the new thinking needed to refresh cultural policy thinking after the "creative turn." In innovation policy studies, a trove of recent publications reframe innovation for a time of virulent tech-backlash and disillusionment with the lures of "disruptive innovation."[33] The two calls for policy renewal, formulated separately, beg to be combined constructively. The Canadian experience of an alternative technological ethos, conceived as intrinsically attuned to the innovation process itself, offers evidence for the value and promise of such a fusion.

Appendix A: List of Interviews Conducted for This Book

Baecker, Ronald M. August 28, 1998, August 17, 2007.

Burtnyk, Nestor. December 15, 1997.

Buxton, Bill. June 24, 1998.

Calvert, Tom. June 14, 2017.

Catmull, Edwin. January 12, 2001.

Clement, Andrew. December 29, 1999.

Davies, Char. June 21, 2018.

Diamond, Sara. March 17, 2017.

Forget, Robert. March 16, 1998.

Freeman, Christopher. November 24, 1998.

Frenkel, Vera. August 3, 2015.

Harrison, John. June 27, 2018.

Hébert, Pierre. September 17, 1999.

Hermanovic, Greg. March 19, 1999.

Jodoin, René. July 3, 1998.

Juneau, Pierre. August 12, 1998.

Kelly, Keith. July 25, 2013.

King, Robin. August 12, 1998.

Knowlton, Ken. February 21, 1999.

Kochanek, Doris. April 27, 1998.

Langlois, Daniel. October 25, 1999.

Liang, Kar. June 22, 1999.

Low, Colin. February 11, 1998.

MacLeod, Douglas. June 27, 2017.

Martin, Eric. July 18, 2000.

Mezei, Leslie. December 27, 1999.

Pardo, René. February 8, 1999.

Petro, Paul. June 6, 2013.

Prévost, Hélène. August 1, 2018.

Pulfer, Ken. December 15, 1997.

Richards, Catherine. February 13, 1998, June 22, 1999.

Roberts, John P. L. April 1, 2016.

Rokeby, David. August 9, 2017.

Schoeler, Peter. October 12, 1999.

Sherman, Tom. May 16, 2017.

Smith, Alvy Ray. April 20, 2000.

Smith, Lynn. June 2, 2002.

Snow, Michael. June 7, 2017.

Tenhaaf, Nell. June 6, 2013.

Truax, Barry. June 1, 2017.

Tulk, Lorne. April 1, 2016.

Tuori, Martin. June 24, 1998.

Wein, Marceli. August 13, 1998.

Westerkamp, Hildegard. June 7, 2017.

Notes

Chapter 1

1. Powe, *A Canada of Light.*

2. McLuhan, *Understanding Media.*

3. Kroker, *Technology and the Canadian Mind*; Darroch, "The Toronto School."

4. See Randolph, "Influencing Machines." After defining the "technological ethos" as an imperative to maximize both the utility of a task and the efficiency with which it is accomplished, Randolph articulates a potential mediating role for the arts: "Art intervenes between technology and fundamental survival not by proving technology so dangerous that it must be extirpated, not by conceding it is so powerful that it is the best medium through which to comprehend our condition, but by constantly embodying the difference between the cultural and the technological realm. Technology can be perceived for all its limitations relative to the cultural realities it denies. If the arts do not insist that technology be perceived as a found object, malleable, revisable, unfinished, culture will have abdicated its power" (52).

5. Charland, "Technological Nationalism."

6. For a consideration of Canadian identity in terms of the contrapuntal "de-hierarchization" of multiple voices, see Heble, "New Contexts of Canadian Criticism"; Waterman, "Improvisation and the Audibility of Difference," provides a critical reading of Canadian multiculturalism from the field of critical improvisation studies, going beyond the "sedative" politics of recognition to draw out the co-creative potential in dissensus.

7. Stark, *The Sense of Dissonance.*

8. Callon, "Variety and Irreversibility in Networks of Technique Conception and Adoption."

9. Raboy, "Communication and Globalization."

10. Michael Ignatieff, quoted in Gopnik, "The Return of the Native," 29.

11. Gagnon and Marchessault, *Reimagining Cinema*.

12. Theall, "Expo 67 as Total Environment," 19–20.

13. Riar, "Expo 67, or the Architecture of Late Modernity," 463, 20.

14. Rutherdale and Miller, "'It's Our Country.'"

15. Kuffert, *A Great Duty*, 231.

16. Our World, "Our World Television Broadcast June 25, 1967." Thanks to Ron Graham for pointing me to this source.

17. Saunders, "Canada 150."

18. Berland, "Nationalism and the Modernist Legacy," 19.

19. Henighan, *The Presumption of Culture*, 12.

20. Keck, "Making Work"; Ostry, *The Cultural Connection*.

21. Ostry, *The Cultural Connection*, 117.

22. Bronson, "The Humiliation of the Bureaucrat."

23. Telecommission, *Instant World (Sans Frontières)*, 27–28.

24. Telecommission, *Instant World (Sans Frontières)*, 29.

25. Snow, *The Two Cultures and the Scientific Revolution*.

26. Raboy and Shtern, *Media Divides*; Dakroury, "Present at the Creation."

27. First chairman of the CRTC, Deputy Minister of Communication, President of the CBC.

28. Pierre Juneau, interview with the author, August 12, 1998.

29. Powe, *Mystic Trudeau*, 120.

30. Smith, *Creative Britain*.

31. Applebaum and Hébert, "Report of the Federal Cultural Policy Review Committee," 51–52.

32. Keith Kelly, interview with the author, July 25, 2013.

33. Tom Sherman, interview with the author, May 16, 2017; Sherman, "Cultural Engineering."

34. Communications Canada, "Communications for the Twenty-First Century."

35. Taylor, "The Twentieth Century in the Rear View Mirror," 80.

36. Isaacson, *The Innovators.*

37. Mitchell, Inouye, and Blumenthal, *Beyond Productivity.*

38. Surveyed in Martin, "The Evolution of Science Policy and Innovation Studies."

39. Freeman, *Long Wave Theory*; Freeman, *The Economics of Industrial Innovation*; Perez, "Technological Revolutions and Techno-Economic Paradigms."

40. Castells, *The Rise of the Network Society.*

41. Gibbons et al., *The New Production of Knowledge*, 110.

42. Freeman and Louçã, *As Time Goes By*, 333–334.

43. Barry, Born, and Weszkalnys, "Logics of Interdisciplinarity." Barry and Born define an "ontological" logic in interdisciplinarity, which is at once generative of new kinds of knowledge challenging traditional disciplinary boundaries and conventions, and which—as is clearly the case in this book—is also entangled with, but not reducible to, separate logics of innovation and accountability.

44. Cox, "The Post-Digital and the Problem of Temporality." Cox theorizes the "post-digital" in relation to overlapping temporalities of successive techno-economic paradigms: he refers to the cultural dynamics of these long waves as a "palimpsest of emergent and residual forms" (154).

45. Knorr-Cetina, *Epistemic Cultures.*

46. Jones and Galison, *Picturing Science, Producing Art*, 2.

47. Galison and Jones, "Factory, Laboratory, Studio."

48. Century, "Pathways to Innovation in Digital Culture"; Century, "Cultural Laboratories." Studio labs were defined in these studies as "hybrid sites where new media technologies are designed and developed in co-evolution with their creative application." See also Orrghen, "Surveying the Literature on Technoscience Art"; Wershler, Emerson, and Parikka, "Lab Space"; Parikka, "The Lab Imaginary."

49. Born, *Rationalizing Culture.*

50. Brown et al., *White Heat, Cold Logic.*

51. McLuhan, "Canada: The Borderline Case."

52. Lundvall, "National Innovation Systems."

53. McLuhan, "Canada: The Borderline Case," 226. McLuhan refers here to Canada as the "Land of the DEW Line," referring to Cold War-era Distant Early Warning systems.

54. Wisnioski, Hintz, and Kleine, *Does America Need More Innovators?*; Godin and Vinck, *Critical Studies of Innovation*; Suchman, "The End of Innovation."

55. Chapman and Sawchuk, "Research-Creation"; Borgdorff, *The Conflict of the Faculties*; Salter, Burri, and Dumit, "Art, Design, and Performance."

56. For instance, the media art histories conference and attendant publications http://www.mediaarthistory.org/.

57. Valuable exceptions to this generalization are Salter, *Entangled*; Higgins and Kahn, *Mainframe Experimentalism*.

58. Becker, "The Work Itself."

59. Abramson and Raboy, "Policy Globalization and the 'Information Society'"; Milz, "Canadian Cultural Policy-Making at a Time of Neoliberal Globalization"; Beale, "Identifying a Policy Hierarchy."

60. Moser, Richards, and Tenhaaf, *Virtual Seminar on the Bioapparatus*.

61. Moser and MacLeod, *Immersed in Technology*.

Chapter 2

1. http://onf-nfb.gc.ca/en/about-the-nfb/organization/mandate/.

2. Hébert, *L'ange et l'automate*; Hébert, "Considérations très personnelles sur l'état du cinéma d'animation"; Bendazzi, *Animation*, vol. 3, chap. 2.

3. Thomas, *Disney Animation*, 85–241.

4. Wells, *Understanding Animation*, 36.

5. Colin Low, interview with the author, February 11, 1998.

6. For background see Boje, "Stories of the Storytelling Organization."

7. Churchill, "An Interview with Evelyn Lambart." By "cel work," Lambart is referring to hand-drawing of characters and scenes on overlapping celluloid sheets, in a hierarchical division of labor where decision making descends from the creative work of the director to story artists, experimental animators, and ultimately "in-between" artists tasked with more humdrum work of filling in the small but subtle differences between keyframes.

8. Downie, "Choreographing the Extended Agent," 20.

9. Rosebush, "The Proceduralist Manifesto," 55–56. For an analysis by an economist of the differences between "humdrum" and "creative" inputs to cultural goods, see Caves, *Creative Industries*.

10. This section is based on four personal interviews with the author: Kar Liang, June 22, 1999; Greg Hermanovic, March 19, 1999; Ronald M. Baecker, August 28, 1998 and August 17, 2007; Ken Knowlton, February 21, 1999.

11. McLaren, "Animated Films," 53.

12. Sifianos, "The Definition of Animation," 62.

13. Seconded to the 1969 Telecommission on the future of telecommunications in Canada, he helped formulate a vision of future multimedia networks. Telecommission, *Instant World (Sans Frontières).*

14. Edwards, *The Closed World.*

15. Sutherland, "Sketchpad."

16. Baecker, "Picture-Driven Animation"; Baecker, "Interactive Computer-Mediated Animation."

17. Sutton, "Stan VanDerBeek's *Poemfields.*"

18. For more on the narrowing of Bell Labs' artistic openness in the 1980s, see Kane, *Chromatic Algorithms.*

19. McWilliams, *Creative Process.*

20. McWilliams, *Creative Process.*

21. Bendazzi, *Animation*, vol. 2, 33.

22. McWilliams, *Creative Process.*

23. McWilliams, *Creative Process.*

24. Martin, "Le mystère d'un cinéma instrumental"; Bastiancich, *Norman McLaren.*

25. He was employed as well by Pierre Juneau to advise him on new images at the Canadian Radio and Telecommunication Commission, Canada's broadcasting regulator.

26. Martin, "Cinéma d'animation et ordinateur."

27. Martin, "Le mystère d'un cinéma instrumental," 28.

28. Bastiancich, *Norman McLaren*, 10.

29. Bastiancich, *Norman McLaren*, 11.

30. Bastiancich, *Norman McLaren*, 28.

31. Bastiancich, *Norman McLaren*, 29.

32. Bastiancich, *Norman McLaren*, 31.

33. Dobson, *The Film Work of Norman McLaren.*

34. Bastiancich, *Norman McLaren*, 34.

35. Hébert, *L'ange et l'automate*, 24.

36. Considered aesthetically as well, in Walter Benjamin's sense that avant-garde art offers a premonition of future formats and technical standards: "One of the foremost tasks of art has always been the creation of a demand which could be fully

satisfied only later. The history of every art form shows critical epochs in which a certain art form aspires to effects which could be fully obtained only with a changed technical standard, that is to say, in a new art form." Benjamin, "The Work of Art in the Age of Mechanical Reproduction," fn. 17.

37. Bertrand and Jean, *Norman McLaren: The Master's Edition*, disk 7. In addition, each film, other than the computer film, is well documented in the Film Board's published technical reports. (Cited in McWilliams, *On the Creative Process*.) The analysis of the breadth of this production, spanning from the late 1950s to the early 1970s, is made possible by the publication in 2006 of comprehensive 7-disk DVD set of McLaren's work, drawing on unreleased material in his archive, including both unreleased films as well as test footage.

38. Churchill, "An Interview with Evelyn Lambart."

39. Quoted in Mollaghan, "An Experiment in Pure Design," 29.

40. McLaren, quoted in McWilliams, *On the Creative Process*, 76.

41. Curtis, "Locating McLaren?," 8.

42. McLaren, quoted in McWilliams, *On the Creative Process*, 76.

43. Mollaghan, "An Experiment in Pure Design," 31.

44. McWilliams, *Creative Process*; Dobson, *The Film Work of Norman McLaren*.

45. Gould, "Where Music and Film Meet."

46. Curtis, "Locating McLaren?," 8.

47. Joseph, *Beyond the Dream Syndicate*.

48. Deleuze, *Cinema 2*, 221.

49. McWilliams, "A God in Volda," 36.

50. McWilliams, "A God in Volda," 36.

51. Davies, "Drawn Sound."

52. Dobson, "'Bump . . . Bup . . . Bup.'"

53. McLaren, quoted in McWilliams, *On the Creative Process*, 91.

54. Dobson, "'Bump . . . Bup . . . Bup,'" 106.

55. Dobson, "'Bump . . . Bup . . . Bup,'" 106.

56. Manovich, *The Language of New Media*.

57. Rosenthal, "Norman McLaren on Pas de Deux."

58. Manning, *Politics of Touch*, 118.

59. Thain, "In the Blink of an Eye," 171.

60. Thain, "In the Blink of an Eye," 168.

61. René Pardo, interview with the author, February 8, 1999.

62. Mezei, "Computer Art."

63. Hébert, *L'ange et l'automate.*

64. Crombie, "Experimental Science and the Rational Artist in Early Modern Europe."

65. Catmull, "Interview"; Catmull and Wallace, *Creativity, Inc.*

66. Catmull and Wallace, *Creativity, Inc.*, 17.

67. Utterson, "A Computer Animated Hand."

68. Smith, "The Stuff of Dreams"; Catmull and Wallace, *Creativity, Inc.*

69. Lévi-Strauss, *The Savage Mind*, 12.

70. Turkle and Papert, "Epistemological Pluralism and the Revaluation of the Concrete," 6.

71. Latour, "Visualisation and Cognition," 2.

72. Latour, "Visualisation and Cognition," 2.

73. Poggioli, *The Theory of the Avant-Garde*, 136–137.

74. Mintzberg and McHugh, "Strategy Formation in an Adhocracy." McGill University management scholar Henry Mintzberg has written on the NFB's management structures during these decades as the epitome of "strategy formation in an adhocracy."

75. Graham, *Canadian Film Technology*. For a development of the studio laboratory construct see Century, "Pathways to Innovation in Digital Culture."

76. Graham, *Canadian Film Technology*, 144.

Chapter 3

1. Patterson, "From the Gun Controller to the Mandala"; Sutton, "Stan VanDerBeek's *Poemfields*"; Johnston, "Models of Code and the Digital Architecture of Time"; Utterson, "Early Visions of Interactivity"; Kane, *Chromatic Algorithms*; Higgins and Kahn, *Mainframe Experimentalism*.

2. Caves, *Creative Industries*, 85–86.

3. Higgins and Kahn, *Mainframe Experimentalism*.

4. Edwards, *The Closed World*.

5. Wein and Burtnyk, "Computer Animation."

6. Throughout this section, primary sources are personal interviews with the author as cited individually.

7. Ken Pulfer, interview with the author, December 15, 1997.

8. Schön, *The Reflective Practitioner.*

9. Pulfer, "Man-Machine Communication in Creative Applications," 1.

10. Ken Pulfer, interview with the author, December 15, 1997.

11. Pulfer, "Man-Machine Communication in Creative Applications," 3.

12. Pulfer, "Man-Machine Communication in Creative Applications," 3.

13. Pulfer, "Man-Machine Communication in Creative Applications," 3.

14. Engelbart and English, "A Research Centre for Augmenting Human Intellect."

15. Hiltzik, *Dealers of Lightning*; Bardini, *Bootstrapping*; Kay, "User Interface."

16. Bill Buxton, interview with the author, June 24, 1998.

17. Nestor Burtnyk, interview with the author, December 15, 1997.

18. McLaren, "Animated Films," 53.

19. René Jodoin, interview with the author, July 3, 1998.

20. René Jodoin, interview with the author, July 3, 1998.

21. Moretti, "Memorandum to René Jodoin."

22. Pierre Hébert, interview with the author, September 17, 1999.

23. Martin, "Cinéma d'animation et ordinateur."

24. Wolf, "A Brief History of Morphing," 90.

25. McWilliams, *Creative Process.*

26. Catmull, "The Problems of Computer-Assisted Animation," 497.

27. Edwin Catmull, interview with the author, January 12, 2001.

28. The "American computer" Foldes refers to was the Scanimate, an analogue system used commercially for television ads and special effects. "Disons que l'ordinateur américain a des yeux et l'ordinateur canadien une main. Les Amérciains ont des impératifs commerciaux, un souci de rentabilité. Les Canadiens du CNRS sont beaucoup plus désintéressés et subventionnent la recherche pure" (my translation above).

29. Foldes, "Interview" (emphasis added).

30. Burtnyk and Wein, "Interactive Skeleton Techniques."

31. Burtnyk and Wein, "Interactive Skeleton Techniques."

32. Wein, "Twenty-Four Years of Computer Graphics at NRC."

33. Steinmueller, "The U.S. Software Industry."

34. Kane, *Chromatic Algorithms*, 173.

35. Mintzberg and McHugh, "Strategy Formation in an Adhocracy."

36. Kochanek, "A Computer System for Smooth Keyframe Animation."

37. Daniel Langlois, interview with the author, October 25, 1999.

38. Deleuze, *Cinema 2*, 221.

39. Daniel Langlois, interview with the author, October 25, 1999.

40. Burnham, "Art and Technology."

41. Hébert, "Les enjeux de l'art à l'ère des machines."

42. René Jodoin, interview with the author, July 3, 1998.

43. Doris Kochanek, interview with the author, April 27, 1998.

44. Parke, "Control Parameterization for Facial Animation."

45. Price, *The Pixar Touch*, 103.

46. Daniel Langlois, interview with the author, October 25, 1999.

47. Kane, *Chromatic Algorithms*.

48. Pierre Hébert, *L'ange et l'automate*, 129 (my translation).

49. Martin Tuori, interview with the author, June 24, 1998; Peter Schoeler, interview with the author, October 12, 1999.

50. Patterson, "From the Gun Controller to the Mandala"; Halpern, *Beautiful Data*.

51. Kane, *Chromatic Algorithms*.

52. Price, *The Pixar Touch*.

53. Youngblood, "Cinema and the Code," 4.

54. Catmull and Wallace, *Creativity, Inc.*; Alvy Ray Smith, interview with the author, April 20, 2000.

Chapter 4

1. Hankins and Silverman, *Instruments and the Imagination*.

2. McLuhan, "Canada: The Borderline Case," 233 (emphasis added).

3. Galloway, *The Interface Effect*, 31, 33, 32.

4. Berland, *North of Empire*, 3.

5. Fourmentraux, "Governing Artistic Innovation."

6. Berland, *North of Empire*, 11.

7. Suchman, *Human-Machine Reconfigurations*.

8. See various discussions from this standpoint in Laurel, *The Art of Human-Computer Interface Design*.

9. Engelbart and English, "A Research Centre for Augmenting Human Intellect"; Bardini, *Bootstrapping*.

10. Greelish, "An Interview with Computing Pioneer Alan Kay."

11. Hughes, *Rescuing Prometheus*; Lenoir, "Virtual Reality Comes of Age"; Buxton, "Interaction at Lincoln Laboratory in the 1960s."

12. Licklider, "Man-Machine Symbiosis."

13. Edwards, *The Closed World*, 266.

14. Lenoir, "Virtual Reality Comes of Age," 229.

15. Except where otherwise noted, sources for this section are interviews of Ronald M. Baecker with the author, August 28, 1998 and August 17, 2007.

16. Norman McLaren, quoted in Sifianos, "The Definition of Animation."

17. Eric Martin, quoted in Baecker, "Genesys," 97.

18. Baecker, "Genesys," 109.

19. Eric Martin, interview with the author, July 18, 2000.

20. Foley et al., *Computer Graphics*.

21. Baecker, "Picture-Driven Animation."

22. Lynn Smith, interview with the author, June 2, 2002.

23. Nelson, *Computer Lib/Dream Machines*.

24. Gotlieb and Borodin, *Social Issues in Computing*.

25. Andrew Clement, interview with the author, December 29, 1999.

26. Leslie Mezei, interview with the author, December 27, 1999.

27. Mezei and Rockman, "The Electronic Computer as an Artist."

28. Wisnioski, "Why MIT Institutionalized the Avant-Garde."

29. Hiltzik, *Dealers of Lightning*.

30. Kay, "The Early History of Smalltalk," 549.

31. Waldrop, *The Dream Machine*.

32. Kay and Goldberg, "Personal Dynamic Media," 38.

33. Ronald M. Baecker, interview with the author, August 2007.

34. Baecker, "A Conversational Extensible System" (emphasis added).

35. Kay, "The Early History of Smalltalk."

36. Smith, "Pygmalion."

37. Quoted in Weibel, *Beyond Art*, 243.

38. Leslie Mezei, interview with the author, December 27, 1999; Bill Buxton, interview with the author, June 24, 1998.

39. Pulfer, "Man-Machine Communication in Creative Applications."

40. Ken Pulfer, interview with the author, December 15, 1997.

41. Pulfer, "Man-Machine Communication in Creative Applications."

42. Holmes and Pender, *Electronic and Experimental Music*, 197–200; Lewin-Richter and Lanza, "On Hugh Le Caine"; Young, *Hugh Lecaine, Pioneer in Electronic Music*.

43. Tal, "An Encounter with Hugh Le Caine," 50.

44. Gell, *Art and Agency*; Lewis, "Too Many Notes," 37; Born, "On Musical Mediation."

45. Boulez, *Passages du vingtième siècle*, 95.

46. Transcribed from a 2012 lecture, https://www.youtube.com/watch?v=ZQJIwjlaPCQ. See also Buxton, "Artists and the Art of the Luthier."

47. Lanier, *Dawn of the New Everything*, 145.

48. Buxton, "My Vision Isn't My Vision."

49. Reichardt, *Cybernetics, Art, and Ideas*, 174ff.

50. Buxton, "Structured Sound Synthesis Project."

51. Buxton et al., "A Microcomputer-Based Conducting System," 8. The authors note that the presentation of their diagram reproduced here as figure 4.2, and concepts underlying it, owed much to conversations with computer music pioneer Max Matthews of Bell Labs.

52. Lee, Buxton, and Smith, "A Multi-Touch Three Dimensional Touch-Sensitive Tablet."

53. NPR radio interview, http://www.npr.org/2011/12/26/144146395/the-touchy-feely-future-of-technology.

54. Baecker and Buxton, *Readings in Human-Computer Interaction: A Multi-Disciplinary Approach*; Baecker et al., *Readings in Human-Computer Interaction: Toward the Year 2000*.

55. Tom Calvert, interview with the author, June 14, 2017.

56. Calvert and Chapman, "Notation of Movement with Computer Assistance."

57. Kittler, *Gramophone, Film, Typewriter*, 3–4.

58. Dufourt, "L'artifice d'écriture dans la musique occidentale," 470 (my translation).

59. Goodman, "Some Notes on Languages of Art."

60. Rotman, "Corporeal or Gesturo-Haptic Writing," 431.

61. Cunningham, *Changes*.

62. Quoted in Schiphorst, "A Case Study of Merce Cunningham's Use of the Lifeforms Computer Choreographic System," 66.

63. Downie, "Choreographing the Extended Agent," 22.

64. Calvert et al., "The Evolution of an Interface for Choreographers."

65. Schiphorst, "A Case Study of Merce Cunningham's Use of the Lifeforms Computer Choreographic System," 92, 94.

66. Sitney, *Visionary Film*, 348.

67. Sitney, *Visionary Film*, 348, quoting Annette Michelson.

68. Snow, *The Collected Writings of Michael Snow*, 53.

69. Snow, *The Collected Writings of Michael Snow*, 63.

70. Snow, *The Collected Writings of Michael Snow*, 63.

71. Snow, *The Collected Writings of Michael Snow*, 56.

72. Testa, "An Axiomatic Cinema," 62.

73. Locke, "Michael Snow's 'La Region Centrale,'" 67.

74. Fleischer, "La cinémachine de Michael Snow," 41.

75. Snow, *The Collected Writings of Michael Snow*, 58.

76. White, "Strangeloves," 51.

77. Snow, *The Collected Writings of Michael Snow*, 89.

78. Hankins and Silverman, *Instruments and the Imagination*, 83ff.

79. Brougher et al., *Visual Music*; Duplaix, *Sons et lumières*.

80. Whitney, *Digital Harmony*.

81. Michael Snow, interview with the author, June 7, 2017.

82. Friis-Hansen, *Visionary Apparatus*, 28.

83. Tomas, *Vertov, Snow, Farocki*.

84. Pickering, *The Cybernetic Brain*, 6.

85. Gagnon, "Faire comme les musiciens." Gagnon provides an insightful reading of diverse media and visual artists working with instruments enabling them to "play" like musicians.

86. Michael Snow, interview with the author, June 7, 2017.

87. Snow, *The Collected Writings of Michael Snow*, 81.

88. Ascott, "Behaviourist Art and the Cybernetic Vision," 39–41.

89. Moores, *Art Creates Change*.

90. Hill, "Photo/Electric Art Show," 29.

91. Quoted in Moser, Richards, and Tenhaaf, *Virtual Seminar on the Bioapparatus*, 70.

92. Rokeby, "The Harmonics of Interaction."

93. Rokeby, "Context-Awareness and Meaning."

94. David Rokeby, interview with the author, August 9, 2017.

95. David Rokeby, interview with the author, August 9, 2017.

96. Rokeby, "Transforming Mirrors."

97. Rokeby, "Transforming Mirrors," 148.

98. Lewis, "Real Time Composing and Improvising Machines."

99. Lewis, "Too Many Notes," 34.

100. Rokeby, "Transforming Mirrors," 145.

101. Rokeby, "Transforming Mirrors," 147.

102. Kwastek, *Aesthetics of Interaction in Digital Art*, 239.

103. Lewis, "Too Many Notes."

104. Penny, "Improvisation and Interaction, Canons and Rules, Emergence and Play," 416.

105. Greelish, "An Interview with Computing Pioneer Alan Kay."

106. Chun, *Programmed Visions*, 60.

107. Chun, *Programmed Visions*, 60.

Chapter 5

1. Gould, "An Argument for Music in the Electronic Age," 231.

2. Gould, "An Argument for Music in the Electronic Age," 231.

3. Gould, "Dialogues on the Prospects of Recording."

4. Gould, "The Prospects of Recording."

5. Gould, "Dialogues on the Prospects of Recording."

6. Galloway, *The Interface Effect*, 32.

7. Gould, "The Prospects of Recording," 347.

8. Oliveros, "Composing with Hugh Le Caine's Special Purpose Tape Recorder."

9. Jones-Imhotep, "Malleability and Machines."

10. Jones-Imhotep, "Malleability and Machines," 305.

11. Gould, "The Prospects of Recording," 347.

12. Gould, "The Prospects of Recording," 353.

13. See Goehr, *The Quest for Voice*, chap. 4, for a discussion of this point.

14. Douglas, "Rightness of Categories," 256, 254.

15. Brougher et al., *Visual Music*.

16. Youngblood, *Expanded Cinema*.

17. Sterne, "Sonic Imaginations," 9–10.

18. Williams, "Acoustic Space."

19. Extensive treatments can be found in Cavell, "Gould, McLuhan, and the Fate of the Acoustic"; Breitsameter, "Schafer's and McLuhan's Listening Paths."

20. Williams, "Acoustic Space," 12–13 (emphasis added).

21. Schafer, *The Tuning of the World*.

22. Schafer, *On Canadian Music*, 47.

23. Johnston, *Radical Campus*, 236.

24. Schafer's standing as one of the initiators of sound studies is contested by some scholars who charge him with a nostalgic romanticism and taking a "prescriptive" position against technologically mediated sound. Kelman, "Rethinking the Soundscape"; Sterne, *The Audible Past*. For a rebuttal to these critiques, written by a scholar who is more sympathetic to the totality of Schafer's contributions, see the

introductory essay by Sabine Breitsameter in Breitsameter and Leonardson, *Ways of Listening, Figures of Thought.*

25. Truax, *Handbook for Acoustic Ecology*, iv.

26. Truax, *Handbook for Acoustic Ecology*, 165.

27. Schafer, *The Tuning of the World*, 208.

28. Schafer, "Acoustic Design," 8.

29. Schafer, "Acoustic Design," 8.

30. Truax, *Handbook for Acoustic Ecology*, 126.

31. Truax, *Handbook for Acoustic Ecology*, 130.

32. Schafer, *String Quartet 2*, 1.

33. Schafer, *Voices of Tyranny, Temples of Silence*. "The *Princess of the Stars* . . . is the most 'Canadian' work I have written, if by Canadian one means something that authentically reflects the habit of living in a place one knows and loves."

34. Schafer's use of indigenous myths has been subject to a justified critique, as when Ellen Waterman observes how "*Patria* presents an essentialized worldview disturbingly out of touch with current Canadian society." Waterman, "*Patria* at the Millennium," 22.

35. Except where otherwise cited, this section is based on a personal interview of Hildegard Westerkamp with the author, June 7, 2017.

36. Westerkamp, "Soundwalking," 49–50.

37. Oliveros, *Sonic Meditations*.

38. Westerkamp, "Linking Soundscape Composition and Acoustic Ecology."

39. Westerkamp, "Acoustic Ecology and the Zone of Silence," 8.

40. Buell, *Proceedings of the First International Conference on Acoustic Ecology*.

41. Except where otherwise cited, this section is based on a personal interview of Barry Truax with the author, June 1, 2017.

42. Truax, *Acoustic Communication*, 3.

43. Truax, "Editorial: Context-Based Composition," 1.

44. Truax, "Editorial: Context-Based Composition," 1.

45. Truax, "Computer Music and Acoustic Communication."

46. Wolff, *Johann Sebastian Bach*.

47. Wolff, *Johann Sebastian Bach*, 306.

48. Truax, "Soundscape, Acoustic Communication and Environmental Sound Composition," 63.

49. Truax, "Real-Time Granular Synthesis with a Digital Signal Processor," 19–20.

50. Helmuth, "Barry Truax's *Riverrun*," 187.

51. Truax, "Real-Time Granular Synthesis with a Digital Signal Processor," 25.

52. Truax, "Real-Time Granular Synthesis with a Digital Signal Processor," 25.

53. Barry Truax, interview by Michael E. Karman, 2009.

54. Truax, "Real-Time Granular Synthesis with a Digital Signal Processor," 19.

55. Breitsameter, "Schafer's and McLuhan's Listening Paths."

56. Hildegard Westerkamp, interview with the author, June 7, 2017.

57. Lewis, "Noise on the Western Front," for an account of this San Francisco Bay Area musical scene.

58. Bartlett, "A Microcomputer-Controlled Synthesizer for Live Performance," 29–30.

59. Fowler, *Electro-Pythagoras*.

60. Described by one of its instructors, at Lewis, "Noise on the Western Front."

61. Scheidt, "Computed Music," 5 (emphasis added).

62. Rosenboom, *Biofeedback and the Arts*.

63. Rosenboom, *Invisible Gold*.

64. Rosenboom, *Invisible Gold*.

65. Rosenboom, "On Being Invisible," 11.

66. Pauline Oliveros, Frederic Rzewski, and George Lewis (presenting early electronic systems) interacted in this workshop with jazz luminaries including Lee Konitz, Ed Blackwell, Sam Rivers, Dave Holland, and Karl Berger. David Lancashire, "Creative Rhythm in the Rockies," *Toronto Globe and Mail*, December 11, 1980.

67. Monahan, "A Half Hour Behind.

68. Robineau, "La musique 'actuelle' québécoise," 105.

69. Lefebvre, "L'underground musical des années 1970," 253.

70. Lefebvre, "L'underground musical des années 1970," 256.

71. Robineau, "La musique 'actuelle' québécoise," 111.

72. Denis, "René Lussier," 32.

73. Hélène Prévost, interview with the author, August 1, 2018.

74. Dhomont, "But Is This Music?," 120.

75. Valiquet, "Animating the Object."

76. Valiquet, "Animating the Object," 388.

77. Garro, "From Sonic Art to Visual Music," 105.

78. Piché, "Music on Image."

79. Normandeau, ". . . Et vers un cinéma pour l'oreille."

80. Oswald, "Plunderphonics."

81. Oswald, "Plunderphonics," 7

82. Oswald, "Bad Relations," 161.

83. Oswald, "Bad Relations," 161.

84. Kahn, "Sound Art, Art, Music."

85. Prévost, "Écoutes urbaines"; Schryer, "Electroacoustic Sound Ecology"; Dumas, "Écologie sonore."

86. Dumas, "Écologie sonore," 48.

87. John P. L. Roberts, interview with the author, April 1, 2016.

88. Lorne Tulk, interview with the author, April 1, 2016.

89. Gould, *The Idea of North.*

90. More recently, the work has also been roundly condemned for the "benevolent racism" it betrays in the sections dealing with the Inuit and First Nations inhabitants of the North. Vallee, "Glenn Gould's *The Idea of North.*" See also Mantere, *The Gould Variations*, 238.

91. Gould, "Radio as Music," 380.

92. John P. L. Roberts, interview with the author, April 1, 2016.

93. Gould, "Radio as Music," 379–380.

94. Arnold Schoenberg was as much a touchstone composer to Gould as J. S. Bach. See Gould, "Arnold Schoenberg."

95. Heble, "New Contexts of Canadian Criticism," 90.

96. Heble, "New Contexts of Canadian Criticism," 90.

97. Berland, *North of Empire*, 112; Said, "On Glenn Gould"; Kingwell, *Glenn Gould*, 136; Howes, "Glenn Gould's Contrapuntal Constitution."

98. Le Moyne worked as a senior advisor to Prime Minister Trudeau from 1969 to 1978, who later appointed him to the Canadian Senate.

99. See Thibault and Hayward, "Jean Le Moyne's *Itinéraire mécanologique*."

100. Gould, *Glenn Gould on the Moog Synthesizer*, based on my own transcription of the program audio. The following quotes are from the same source.

101. Steve Tulk, Gould's recording technician, explained to me that these early contrapuntal radio pieces were made by running parallel tape recorders in single channels into a mixer while he controlled levels. Lorne Tulk, interview with the author, April 1, 2016.

102. Simondon, "Gilbert Simondon entretien sur la mécanologie," 125.

103. Simondon, *On the Mode of Existence of Technical Objects*.

104. Simondon, "Gilbert Simondon entretien sur la mécanologie," 126–128.

105. Gould, "Rubinstein," 289–290.

106. Gould, *Glenn Gould on the Moog Synthesizer* (emphasis added).

107. Gould, *Glenn Gould on the Moog Synthesizer*.

108. Kingwell, *Glenn Gould*, 168.

109. Payzant, *Glenn Gould*, 136–139.

110. Gould, *A Glenn Gould Fantasy*.

111. Nattiez, "Gould Out of Time," for a musicological angle on Gould's mental health; Ostwald, *Glenn Gould*, for one from the perspective of a physician and friend of Gould's.

112. Nowotny, *Insatiable Curiosity*.

113. Théberge, *Any Sound You Can Imagine*.

114. Loy, "Musicians Make a Standard."

115. Théberge, *Any Sound You Can Imagine*, 225.

116. http://improvisationinstitute.ca/.

Chapter 6

1. Canada's Public Archives have forty-two letters from Marshall McLuhan to Prime Minister Pierre Trudeau; forty-four personally written replies from the Prime Minister to McLuhan. The two also met and dined personally on a number of occasions. McLuhan, *Letters of Marshall McLuhan*, 350.

2. Raboy and Shtern, *Media Divides*. See the discussion of *Instant World* in chapter 1.

3. McLuhan, *Letters of Marshall McLuhan*, 365.

4. McLuhan, *Letters of Marshall McLuhan*, 427.

5. McLuhan, *Understanding Media*; McLuhan, "Media as Art Forms."

6. McLuhan, "Media as Art Forms," 7.

7. McLuhan, "Media as Art Forms," 9.

8. McLuhan, *Letters of Marshall McLuhan*, 427.

9. According to Kar Liang in an interview with the author, June 22, 1999.

10. Gleick, *The Information*.

11. Licklider, "A Picture Is Worth a Thousand Words," 619.

12. Sterne, *MP3*, 3–4.

13. Sterne, "Compression: A Loose History," 35.

14. Licklider, "A Picture Is Worth a Thousand Words," 619.

15. Thompson, "Computers and Visual Literacy."

16. Kac, *Telepresence and Bio Art*, 50–51.

17. Bull, "Dictation," 132.

18. Arnold and Henry, *Traffic*.

19. Alloway, "Network."

20. Quoted in Low, "Connections/Disconnections," 103.

21. For the concept of "imaginary media," see Kluitenberg, *Book of Imaginary Media*.

22. Robertson, *Policy Matters*, 2, quoting *Artnews* magazine, 1977. More broadly, Khonsary and Podesva, *Institutions by Artists*; Detterer and Nannucci, *Artist-Run Spaces*.

23. Nemiroff, "A History of Artist-Run Spaces in Canada," 263.

24. Bronson, "The Humiliation of the Bureaucrat," 167.

25. McKinnie, "Bees, Horseshoes, and Puppets for the Elderly," 428–429.

26. Nemiroff, "A History of Artist-Run Spaces in Canada," 269.

27. Burgess and De Rosa, "The Distinct Role of Artist-Run Centres in the Canadian Visual Arts Ecology," 16–17.

28. Bonin, "Here, Bad News Always Arrives Too Late," 66.

29. Gaudreau, "Robert Filiou."

30. Quoted in Robertson, *Policy Matters*, 16.

31. Bull, "Dictation," 128.

32. Lewis, "The Living Museums Data Network," 145.

33. Lewis, "The Living Museums Data Network," 140.

34. Bartlett, "Speed, Mechanization, Convenience and Accessibility," 88.

35. Nemiroff, "Par-Al-Lel," 19.

36. Davis, "Sat-Tel-Comp"; Turner, "Sat-Tel-Comp"; Jarvis, "Sat-Tel-Comp."

37. Klüver, Martin, and Rose, *Pavilion*.

38. Davis, "Sat-Tel-Comp."

39. Turner, "Sat-Tel-Comp."

40. Turner, "Sat-Tel-Comp."

41. A San Francisco conference co-organized by Bartlett on Artist's Use of Telecommunications included Douglas Davis (New York), Willoughby Sharp and Liza Beár (New York), John Southworth (Hawaii), Aldo Tambellini (M.I.T.), Michael Goldberg (Tsukuba University, Japan), Judith Doyle, Gene Youngblood, Hank Bull and Kate Craig (Vancouver), Glenn Lewis (Vancouver), Robert Adrian X (Vienna), Richard Kriesche (Vienna), Helmut Mark (Vienna) and Norman White (Toronto). (Turner, "Sat-Tel-Comp.")

42. Jarvis, "Sat-Tel-Comp."

43. Shaw, "Expanded Consciousness and Company Types."

44. Bronson, "Artist-Initiated Activity in Canada," 10.

45. Nemiroff, "A History of Artist-Run Spaces in Canada"; Wallace, "A Particular History."

46. http://kunstradio.at/HISTORY/TCOM/WC/wc-main.html.

47. Adrian, "Art and Telecommunications 1979–1986," 239.

48. Sponsored by I. P. Sharp Associates, a business services company supporting these artist exchanges gratis.

49. Adrian, "Art and Telecommunications 1979–1986," 239.

50. Adrian, "Communicating," 79.

51. Ascott, "Art and Telematics," 26.

52. Adrian, "Art and Telecommunications 1979–1986," 239.

53. Bull, "Dictation," 132.

54. Vera Frenkel, interview with the author, August 3, 2015 (source for this section, when not indicated otherwise).

55. Frenkel, "Herman and Herbert," 3. Frenkel's typescript is extended from a presentation at the conference Konfigurationen: Zwischen Kunst und Medien (Configurations: Between Art and Media), Kassel, September 4, 1997, and later, somewhat revised, at the McLuhan Program Coach House Festival, Toronto, October 21, 1998, and at the Center for Gender and Intercultural Studies, SUNY Buffalo, March 23, 1999.

56. Langill, "Vera Frenkel's *String Games* and the Germination of Media Art in Canada," 374–375.

57. Not incidentally, Vera Frenkel's angle on communications technology is at once pragmatic and diffident, well indicated in a witty essay directly addressing the ideas of Marshall McLuhan, set in counterpoint to his Toronto contemporary Northrop Frye. Taking the former's globally circulating memes on media—"the notion of a global village, media as extensions of the body, each new medium cannibalizing its most recent predecessor, the medium as message"—as plausible yet self-evident, Frenkel herself took a more measured stance, one again marked by her self-perception as an outsider and by a sympathy for Frye's deep-time consideration of mythic patterning and archetypes. For Frenkel, "the McLuhan phenomenon was regarded as suspect in his hometown and still is in some quarters. We left it to the Americans and the Europeans to return him to us with vine-leaves in his hair." Frenkel, "Herman and Herbert," 2.

58. Tuer, "Beyond the New Media Frame."

59. Langill, "Vera Frenkel's *String Games* and the Germination of Media Art in Canada." Langill cites widely read texts by Shanken, Grau, Popper, and Paul as scholars who have overlooked Frenkel's precedence over numerous (mainly male) artists in the applications of telecommunications.

60. Frenkel quoted in Tuer, "Beyond the New Media Frame," 46.

61. Ryle, "Improvisation," 77. See also Lewis, "Improvising Tomorrow's Bodies."

62. Ascott, "*Gesamtdatenwerk*," 85–87.

63. Sterne, "Compression," 47.

64. Sterne, "Compression," 34.

65. Licklider, "A Picture Is Worth a Thousand Words," 619.

66. Sawchuk et al., "An Interactive Image Communication System"; Bown and Sawchuk, "Interactive Graphics Applied to Symbol Communication."

67. Bliss, *Semantography (Blissymbolics)*.

68. Eco, *The Search for the Perfect Language*.

69. Okrent, *In the Land of Invented Languages*, 162.

70. Umberto Eco dismissed Bliss symbols as being only a limited language, "unable to express new ideas because every element of expression always corresponds to a predetermined element of content, known in advance" (Eco, *The Search for the Perfect Language*, 168). However, his dismissal appears to be belied by the reported experience of the children at the OCCC in devising new concepts and expressions. Radiolab, "Mr. Bliss."

71. Bown and Sawchuk, "Interactive Graphics Applied to Symbol Communication," 203–204.

72. Thompson, "Visual Literacy and the Time-Varying Icon."

73. Quoted in Charland, "Technological Nationalism," 220.

74. Government of Canada Department of Communications and National Film Board, *Telidon*.

75. Government of Canada Department of Communications and National Film Board, *Telidon*.

76. Fleming and Frezza, "NAPLPS."

77. Fleming and Frezza, "NAPLPS," 210.

78. Lyman, *Canada's Video Revolution*; Gillies, "Technological Determinism in Canadian Telecommunications."

79. Babe, *Telecommunications in Canada*.

80. Mosco, *Pushbutton Fantasies*.

81. Guédon, "Norme ou système technique?"

82. Feenberg, *Alternative Modernity*.

83. Coll, "Yesterday's Adventures with Today's Technology."

84. Sterling, "Pioneering Risk."

85. Langill, "The Telidon Trials"; Devon, "Interactivity and the Popular Support for Telidon."

86. Sawchuk, "Media Arts Revisited."

87. Robin King, interview with the author, August 12, 1998; Bill Buxton, interview with the author, June 24, 1998.

88. Cartier, "Making Pages." David Godfrey will appear in the next chapter of this book, as the founding chair of the ill-fated CIRCE, a proposed federal center for research on culture and technology that was meant to be founded in Montréal in 1991 but was canceled for political reasons.

89. Cartier, "Making Pages," 138.

90. Cartier, "Making Pages," 141.

91. Thompson, "Visual Literacy and the Time-Varying Icon," 304–305.

92. Thompson, "Visual Literacy and the Time-Varying Icon," 305. This aspect of the Telidon imaginary is precisely what attracted the interest of the system's central political sponsor, Pierre Juneau. Juneau had previously directed Canadian content regulation, was later a CBC president, and in the 1970s served as deputy minister in the Department of Communications and so was instrumental in underwriting the Government's investment in Telidon. In a 1998 interview Juneau explained how he was at the time "fascinated by the idea of making complex notions more easily understandable through visual means. Not in a Disney way, with dazzling effects, but with diagrams, maps, and more. I thought Telidon could be a useful tool to summarize, diagrammize, simplify complex notions." Pierre Juneau, interview with the author, August 12, 1998.

93. Perry, "Computerese," 45.

94. Perry, "Computerese," 45.

95. Shea, "Dead Medium."

96. Paul Petro, interview with the author, June 6, 2013.

97. http://www.vtape.org/video?vi=1819. See also the May 2018 online exhibition of Canadian artists' work in Telidon organized by the InterAccess gallery in Toronto: https://interaccess.org/exhibition/telidon. The full results of a digital restoration project on Telidon art will be exhibited in Spring 2023.

98. Shulgin, "Art, Power, and Communication"; Nell Tenhaaf, interview with the author, June 6, 2013.

99. Sawchuk, "An Index of Power."

100. Sawchuk, "An Index of Power"; Langill, "The Telidon Trials."

101. Kac, *Telepresence and Bio Art.*

102. Hui and Broeckmann, "Introduction," 16: "The group included philosophers and social scientists such as Jacques Derrida, Bruno Latour, François Chatelet, Christine Buci-Glucksmann, Philippe Lacoue-Labarthe, Isabelle Stengers and Dan Sperber."

103. Telecommission, *Instant World (Sans Frontières)*; see chapter 1 above.

104. Adrian, "Art and Telecommunications 1979–1986," 239.

105. See Burgess and De Rosa, "The Distinct Role of Artist-Run Centres in the Canadian Visual Arts Ecology."

106. Sherman, "Cultural Engineering."

107. Meanwhile, alternative creative scenes have explored the expressive capacity of different resolutions and compression standards, as evidenced by the ongoing interest in various kinds of low-res and retro-aesthetics, including 1990s net art, glitch art and music, 8-bit games, and teletext art festivals.

Chapter 7

1. Now the Banff Centre for Arts and Creativity, which states on its webpage that the Centre is located on the lands of Treaty 7 territory: "We acknowledge the past, present, and future generations of Stoney Nakoda, Blackfoot, and Tsuut'ina Nations who help us steward this land, as well as honour and celebrate this place."

2. Day, "The Banff School of Fine Arts."

3. Bawtree, "Inter-Arts at Banff."

4. Augaitis, *Siting Technology*; Fischer and Falk, *The Event Horizon*; Lander and Lexier, *Sound by Artists*; Augaitis and Lander, *Radio Rethink*.

5. Banff Centre, "A Plan for Implementing Media Arts Programs."

6. Barry and Born, *Interdisciplinarity*. Their analysis takes aim at a corporatist, narrowly instrumental conceptualization of interdisciplinary collaboration, arguing for a sharper differentiation between distinct "logics" of justification for mixing and crossing between disciplines, and also between various "modes" or methodologies of interdisciplinarity. Innovation is one such logic; another is accountability—a demand for social accounting for public or other investments. A third logic, most relevant for my present consideration of the hybridized fields of experimental art and technology, is that of ontology: the impetus to create entirely new kinds of objects not imaginable within a single discipline's purview. In terms of modes of interdisciplinary research, the "agonistic-antagonistic mode . . . is conceived neither as a synthesis nor in terms of a disciplinary division of labour, but as driven by an agonistic or antagonistic relation to existing forms of disciplinary knowledge and practice. Here, interdisciplinarity springs from a self-conscious dialogue with, criticism of or opposition to the intellectual, ethical or political limits of established disciplines or the status of academic research in general" (29).

7. Pickering, "The Mangle of Practice."

8. Banff Centre, "A Plan for Implementing Media Arts Programs."

9. Richards, "Virtual Worlds, Digital Images" (emphasis added).

10. Richards, "Body in Ruins," 263–270.

11. Moser, Richards, and Tenhaaf, *Virtual Seminar on the Bioapparatus*. For these authors the Bioapparatus "combines an understanding of particular philosophies of technology with theories about the technological apparatus, the technologized body, and the new biology" (5).

12. Based on Catherine Richards, interview with the author, June 22, 1999; Langill, "Shifting Polarities."

13. Philippe Quéau, email to the author, June 30, 2000.

14. Catherine Richards, interview with the author, June 22, 1999.

15. Nell Tenhaaf, "Program of Work," Banff Centre Archives, Media Arts, Box 1995-112.

16. Tuori and Century, "As Real as What?"

17. Godfrey, "Statement on the Canadian Institute for Research on Cultural Enterprise."

18. Banff Centre, "Art and Virtual Environments."

19. Moser and MacLeod, *Immersed in Technology*.

20. Moser, Richards, and Tenhaaf, *Virtual Seminar on the Bioapparatus*.

21. Adapted from the list compiled in Randolph, "A City for Bachelors," 153–154.

22. Tomas, "Reflections beyond the Interface of Theory and Practice," 117.

23. Robinett and Naimark, "Artists Explore Virtual Reality," 248.

24. Banff Centre Archives Media Arts, Box 1993.62.

25. Banff Centre Archives Media Arts, Box 1993.62.

26. Moser, Richards, and Tenhaaf, *Virtual Seminar on the Bioapparatus*, 28.

27. Moser, Richards, and Tenhaaf, *Virtual Seminar on the Bioapparatus*, 33, 65.

28. Moser, Richards, and Tenhaaf, *Virtual Seminar on the Bioapparatus*, 95, 103.

29. Leler, "Actor-Based Simulation + Linda =Virtual Environments."

30. Leler, "Actor-Based Simulation + Linda =Virtual Environments."

31. Leler, "Actor-Based Simulation + Linda =Virtual Environments."

32. Banff Centre, "Art and Virtual Environments."

33. For further analysis see Century and Bardini, "Towards a Transformative Set-Up."

34. For a recent review of musically driven virtual and extended realities, see Turchet, Hamilton, and Çamci, "Music in Extended Realities." For a more speculative vision of VR world building as a blend of jazz improvisation and real-time software coding, see Lanier, *Dawn of the New Everything*, 319.

35. Yuxweluptun, "Inherent Rights, Vision Rights," 316.

36. Yuxweluptun, "Inherent Rights, Vision Rights," 316.

37. Nemiroff, Houle, and Townsend-Galt, *Land, Spirit, Power*, 12–13.

38. King, "In/Consequential Relationships."

39. www.catherinerichards.ca.

40. Catherine Richards, "Body Schema," March 1991, Banff Centre Archives, Media Arts, Box 1995-112.

41. Banff Centre Archives, Media Arts, Box 1995-112.

42. Dyson, *Sounding New Media*, 162–163.

43. Dyson, *Sounding New Media*, 162–163.

44. Rothman, "Are We Already Living in Virtual Reality?" Described here are experiments inducing body boundary illusions, similar in concept to those proposed fictively by Richards, like the sensation of phantom limbs.

45. Moser and MacLeod, *Immersed in Technology*, ix.

46. Heim, *Virtual Realism*.

47. Huhtamo, "Time Traveling in the Gallery."

48. Laurel, "Placeholder: Notes on Process."

49. Moser and MacLeod, *Immersed in Technology*, 295.

50. Moser and MacLeod, *Immersed in Technology*, 291.

51. Moser and MacLeod, *Immersed in Technology*, 292.

52. Randolph, "A City for Bachelors," 173.

53. Randolph, "A City for Bachelors," 173.

54. Lozano-Hemmer, "Perverting Technological Correctness."

55. Lewis, "Toni Dove's Nonmodern Ontologies."

56. Benedikt, *Cyberspace*.

57. Moser and MacLeod, *Immersed in Technology*, 306.

58. Moser and MacLeod, *Immersed in Technology*, 306.

59. Hansen, *New Philosophy for New Media*, 177.

60. https://painstudieslab.com/.

61. Letter from Carol Phillips to the Centre for Information Technology Innovation, Banff Centre Archives, Media Arts, Box 1993.62.

62. Char Davies, interview with the author, June 21, 2018.

63. Davies, "Letter to Lorne Falk."

64. John Harrison, interview with the author, June 27, 2018.

65. Munster, *Materializing New Media*; Hansen, *Bodies in Code*.

66. Grau, *Virtual Art.*

67. Dyson, *Sounding New Media.*

68. Bureau of Inverse Technology, "Osmose by Char Davies."

69. Dyson, *Sounding New Media*, 109.

70. Lunenfeld, "Cyborg Economics"; Gilpin and Falk, "Demo Aesthetics."

71. Jha, "Computer-Generated Realities Are Becoming Ubiquitous."

72. Kaganskiy, "Our Sensorium."

Chapter 8

1. Powe, *A Canada of Light*, 36.

2. Powe, *A Canada of Light*, 104, 145.

3. Star, "This Is Not a Boundary Object," 602.

4. McLuhan, "Canada as Counter-Environment."

5. Hobsbawm, *The Age of Extremes.*

6. Purcell, "Digital Media and the Primacy of Content," 9.

7. Purcell, "Digital Media and the Primacy of Content," 13.

8. Purcell, "Digital Media and the Primacy of Content," 9.

9. Smith, *Creative Britain.*

10. Caves, *Creative Industries*; Florida, *The Rise of the Creative Class.*

11. Mitchell, Inouye, and Blumenthal, *Beyond Productivity.*

12. Cunningham, "Creative Enterprises," 292 (emphasis added).

13. Schlesinger, "The Creative Economy."

14. Bliss, "Privatizing the Mind," 14.

15. Taylor, "Can Canada Survive the Charter," 446.

16. Taylor, "Deep Diversity and the Future of Canada," 29.

17. Raboy, "Pas de deux."

18. Beale, "Identifying a Policy Hierarchy," 80. Beale observes that "cultural sovereignty" in the Canadian context goes back to the founding of public broadcasting as "a cultural defense system against American transborder radio."

19. McLuhan, *Letters of Marshall McLuhan*, 365.

20. Abramson and Raboy, "Policy Globalization and the 'Information Society.'"

21. http://grand-nce.ca/.

22. Fourmentraux, "Linking Art and Sciences," 148.

23. Fourmentraux, "Linking Art and Sciences," 140.

24. Banff Centre Archives, Box 1993-62.

25. Cook and Diamond, *Euphoria and Dystopia*.

26. SSHRC research-creation: http://www.sshrc-crsh.gc.ca/funding-financement/apply-demande/background-renseignements/preparing_research_creation_application_idg-preparer_l_application_recherche-creation_sds-eng.aspx.

27. Lowry, "Props to Bad Artists."

28. Borgdorff, *The Conflict of the Faculties*. See in particular chapter 7, "The Production of Knowledge in Artistic Research."

29. Salter, Burri, and Dumit, "Art, Design, and Performance," 153.

30. Panofsky, "Artist, Scientist, Genius," 182.

31. Andersen and Cox, "Post-digital Research—Editorial."

32. Schlesinger, "The Creative Economy."

33. Godin and Vinck, *Critical Studies of Innovation*; Suchman and Bishop, "Problematizing 'Innovation' as a Critical Project"; Wisnioski, Hintz, and Kleine, *Does America Need More Innovators?*; Nowotny, "Beyond Innovation"; Lepore, "The Disruption Machine."

Works Cited

Abramson, Bram Dov, and Marc Raboy. "Policy Globalization and the 'Information Society': A View from Canada." *Telecommunications Policy* 23, no. 10–11 (1999): 775–791.

Adrian, Robert. "Art and Telecommunications 1979–1986: The Pioneer Years" (1995). In *Art and Electronic Media*, edited by Edward A. Shanken. London: Phaidon Press, 2009.

Adrian, Robert. "Communicating." In *Art Telecommunication*, edited by Heidi Grundmann. Vancouver: Western Front; Vienna: BLIX, 1984.

Alloway, Lawrence. "Network: The Art World Described as a System." *Art Forum* 9, no. 11 (1972).

Andersen, Christian Ulrik, and Geoff Cox. "Post-digital Research—Editorial." *APRJA. A Peer-Reviewed Journal About Post-digital Research* (2014).

Applebaum, Louis, and Jacques Hébert. "Report of the Federal Cultural Policy Review Committee." Ottawa: Department of Communications, 1982.

Arnold, Grant, and Karen Henry. *Traffic: Conceptual Art in Canada, 1965–1980*. Vancouver: Vancouver Art Gallery; Edmonton: Art Gallery of Alberta; Toronto: Justina M. Barnicke Gallery; Montréal: Leonard and Bina Ellen Art Gallery; Halifax: Halifax, INK, 2012.

Ascott, Roy. "Art and Telematics." In *Art Telecommunication*, edited by Heidi Grundmann. Vancouver: Western Front; Vienna: BLIX, 1984.

Ascott, Roy. "Behaviourist Art and the Cybernetic Vision" (1967). In *Telematic Embrace: Visionary Theories of Art, Technology, and Consciousness*, edited by Edward Shanken. Berkeley: University of California Press, 2003.

Ascott, Roy. "*Gesamtdatenwerk*: Connectivity, Transformation and Transcendence" (1989). In *Ars Electronica: Facing the Future, a Survey of Two Decades*, edited by Timothy Druckrey with Ars Electronica. Cambridge, MA: MIT Press, 1998.

Augaitis, Daina. *Siting Technology*. Banff: Walter Phillips Gallery, 1988.

Augaitis, Daina, and Dan Lander, eds. *Radio Rethink: Art, Sound and Transmission.* Banff: Walter Phillips Gallery, 1994.

Babe, Robert. *Telecommunications in Canada: Technology, Industry, and Government.* Toronto: University of Toronto Press, 1990.

Baecker, Ronald M. "A Conversational Extensible System for the Animation of Shaded Images." Paper presented at SIGGRAPH 76, Philadelphia, 1976.

Baecker, Ronald M. "Genesys: Interactive Computer Mediated Animation." In *Computer Animation*, edited by John Halas. London: Focal Press, 1974.

Baecker, Ronald M. "Interactive Computer-Mediated Animation." MIT, 1969.

Baecker, Ronald M. "Picture-Driven Animation." Paper presented at the Proceedings of 1969 Spring Joint Computer Conference, 1969.

Baecker, Ronald M., and William A. S. Buxton. *Readings in Human-Computer Interaction: A Multi-Disciplinary Approach.* Los Altos, CA: Morgan Kaufmann, 1987.

Baecker, Ronald M., Jonathan Grudin, William A. S. Buxton, and Saul Greenberg. *Readings in Human-Computer Interaction: Toward the Year 2000.* San Francisco: Morgan Kaufmann, 1995.

Banff Centre. "Art and Virtual Environments: A Research and Development Project." Banff Centre Archive, Box AA27.12, 1991.

Banff Centre. "A Plan for Implementing Media Arts Programs in the Jeanne and Peter Lougheed Building, 1989–92." Banff Centre Archive, Box AA27.12, 1989.

Bardini, Thierry. *Bootstrapping: Douglas Engelbart, Coevolution, and the Origins of Personal Computing.* Palo Alto: Stanford University Press, 2000.

Barry, Andrew, Georgina Born, and Gisa Weszkalnys. "Logics of Interdisciplinarity." *Economy and Society* 37, no. 1 (2008): 20–49.

Barry, Andrew, and Georgina Born. *Interdisciplinarity: Reconfigurations of the Social and Natural Sciences.* Abingdon, UK: Routledge, 2013.

Bartlett, Bill. "Speed, Mechanization, Convenience and Accessibility." In "Spaces by Artists," edited by Tanya Rosenberg, *Parallelogram* 3 (1979): 86–88.

Bartlett, Martin. "A Microcomputer-Controlled Synthesizer for Live Performance." *Computer Music Journal* 3, no. 1 (1979): 29–33.

Bastiancich, Alfio. *Norman McLaren. Précurseur des nouvelles images.* Translated by Marlène di Stefano from the Italian. Paris: Dreamland éditeur, 1997 (1981).

Bawtree, Michael. "Inter-Arts at Banff: A Review of Inter-Arts Activity and Thought since 1850 and a Sketch for Banff's Year-Round Programs." Unpublished report, Banff Centre Archives, AA35b.16, 1978.

Beale, Alison. "Identifying a Policy Hierarchy: Communication Policy, Media Industries, and Globalization." In *Global Culture: Media, Arts, Policy, and Globalization*, edited by Nobuko Kawashima, Diana Crane, and Ken'ichi Kawasaki. London: Routledge, 2016.

Becker, Howard S. "The Work Itself." In *Art from Start to Finish: Jazz, Painting, Writing, and Other Improvisations*, edited by Howard S. Becker, Robert R. Faulkner, and Barbara Kirshenblatt-Gimblett. Chicago: University of Chicago Press, 2006.

Bendazzi, Giannalberto. *Animation: A World History*. 3 vols. Boca Raton, FL: Taylor & Francis Group, 2016.

Benedikt, Michael. *Cyberspace: First Steps*. Cambridge, MA: MIT Press, 1991.

Benjamin, Walter. "The Work of Art in the Age of Mechanical Reproduction" (1936). In *Illuminations*, edited by Hannah Arendt, 217–251. New York: Schocken, 1969.

Berland, Jody. "Nationalism and the Modernist Legacy: Dialogues with Innis." In *Capital Culture: A Reader on Modernist Legacies, State Institutions, and the Value(s) of Art*, edited by Jody Berland and Shelley Hornstein. Montréal: McGill-Queen's University Press, 2000.

Berland, Jody. *North of Empire: Essays on the Cultural Technologies of Space*. Durham: Duke University Press, 2009.

Bertrand, Marc, and Marcel Jean. *Norman McLaren: The Master's Edition*. 7 DVDs. Montréal: National Film Board, 2006.

Bliss, Charles Kasiel. *Semantography (Blissymbolics): A Simple System of 100 Logical Pictorial Symbols, Which Can Be Operated and Read Like 1+2=3 in All Languages*. 2d enl. ed. Sydney: Semantography (Blissymbolics) Publications, 1965.

Bliss, Michael. "Privatizing the Mind: The Sundering of Canadian History, the Sundering of Canada." *Journal of Canadian Studies* 26, no. 4 (1992): 5–17.

Boje, David M. "Stories of the Storytelling Organization: A Postmodern Analysis of Disney." *Academy of Management Journal* 38, no. 4 (1995): 997–1035.

Bonin, Vincent. "Here, Bad News Always Arrives Too Late: Institutional Critique in Canada (1967–2012)." In *Institutions by Artists*, edited by Jeff Khonsary and Kristina Lee Podesva, vol. 1. Vancouver: Fillip Editions, Folio C, 2012.

Borgdorff, Henk. *The Conflict of the Faculties: Perspectives on Artistic Research and Academia*. Leiden: Leiden University Press, 2012.

Born, Georgina. "On Musical Mediation: Ontology, Technology and Creativity." *Twentieth-Century Music* 2, no. 1 (2005).

Born, Georgina. *Rationalizing Culture: IRCAM, Boulez, and the Institutionalization of the Musical Avant-Garde*. Berkeley: University of California Press, 1995.

Boulez, Pierre. *Passages du vingtième siècle.* Paris: Centre Georges Pompidou, 1977.

Bown, H. G., and W. Sawchuk. "Interactive Graphics Applied to Symbol Communication for Non-Speaking Children." *Computers and Graphics* 2 (1977).

Breitsameter, Sabine. "Schafer's and McLuhan's Listening Paths: Convergences, Crossings and Diversions." *Soundscape: The Journal of Acoustic Ecology* 11, no. 1 (2011).

Breitsameter, Sabine, and Eric Leonardson, eds. *Ways of Listening, Figures of Thought: A Festschrift for R. Murray Schafer on the Occasion of His 80th Birthday.* Darmstadt: Soundscape & Environmental Medialab, Hochschule Darmstadt, 2013.

Bronson, A. A. "Artist-Initiated Activity in Canada." In *From Sea to Shining Sea,* edited by A. A. Bronson and Peggy Gale with René Blouin, Glenn Lewis. Toronto: Power Plant, 1987.

Bronson, A. A. "The Humiliation of the Bureaucrat: Artist-Run Spaces as Museums by Artists." In *From Sea to Shining Sea,* edited by A. A. Bronson and Peggy Gale with René Blouin, Glenn Lewis. Toronto: Power Plant, 1987.

Brougher, Kerry, et al. *Visual Music: Synaesthesia in Art and Music since 1900.* New York: Thames & Hudson; Washington, DC: Hirshhorn Museum and Sculpture Garden; Los Angeles: LACMA, 2005.

Brown, Paul, Charlie Gere, Nicholas Lambert, and Catherine Mason. *White Heat, Cold Logic: British Computer Art 1960–1980.* Cambridge, MA: MIT Press, 2008.

Buell, Tim, ed. *Proceedings of the First International Conference on Acoustic Ecology.* 2 vols. Banff: Banff Centre, 1993.

Bull, Hank. "Dictation: A Canadian Perspective on the History of Telematic Art." In *Social Media Archeology and Poetics,* edited by Judy Malloy. Cambridge, MA: MIT Press, 2016.

Bureau of Inverse Technology (BIT). "Osmose by Char Davies." *Mute* 1, no. 3 (1995).

Burgess, Marilyn, and Maria De Rosa. "The Distinct Role of Artist-Run Centres in the Canadian Visual Arts Ecology." Montréal: MDR Burgess Consultants, 2011.

Burnham, Jack. "Art and Technology: The Panacea That Failed." In *The Myths of Information: Technology and Postindustrial Culture,* edited by Kathleen Woodward. Madison, WI: Coda Press, 1980.

Burtnyk, Nestor, and Marceli Wein. "Interactive Skeleton Techniques for Enhancing Motion Dynamics in Key Frame Animation." *ACM Communications* 19, no. 10 (1976).

Buxton, Bill. "Artists and the Art of the Luthier." *Computer Graphics: The SIGGRAPH Quarterly* 31, no. 1 (1997): 10–11.

Buxton, Bill. "Interaction at Lincoln Laboratory in the 1960s: Looking Forward—Looking Back." Paper presented at the CHI 2005, Portland, 2005.

Buxton, Bill. "My Vision Isn't My Vision: Making a Career out of Getting Back to Where I Started." In *HCI Remixed: Reflections on Works That Have Influenced the HCI Community*, edited by Thomas Erickson and David McDonald. Cambridge, MA: MIT Press, 2008.

Buxton, Bill. "Structured Sound Synthesis Project (SSSP)." http://www.billbuxton.com/buxtonSSSPVideos.html.

Buxton, W., W. Reeves, G. Fedorkow, K. C. Smith, and R. Baecker. "A Microcomputer-Based Conducting System." *Computer Music Journal* 4, no. 1 (1980): 8–21.

Callon, Michel. "Variety and Irreversibility in Networks of Technique Conception and Adoption." In *Technology and the Wealth of Nations: The Dynamics of Constructed Advantage*, edited by Dominique Foray and Christopher Freeman. London: Pinter, with OECD, 1993.

Calvert, T. W., A. Bruderlin, S. Mah, T. Schiphorst, and C. Welman. "The Evolution of an Interface for Choreographers." Paper presented at the INTERCHI Conference, Amsterdam, 1993.

Calvert, Thomas W, and John Chapman. "Notation of Movement with Computer Assistance." Paper presented at the Proceedings of the 1978 ACM Annual Conference, vol. 2, 1978.

Cartier, Michel. "Making Pages." In *The Telidon Book*, edited by David Godfrey and Ernest Chang. Victoria: Softwords, 1983.

Castells, Manuel. *The Rise of the Network Society*. Malden, MA: Blackwell, 1996.

Catmull, Edwin. "Interview." *Animation World Magazine*, 2000.

Catmull, Edwin. "The Problems of Computer-Assisted Animation." *Computer Graphics* 12, no. 3 (1978): 495–500.

Catmull, Edwin E., and Amy Wallace. *Creativity, Inc.: Overcoming the Unseen Forces That Stand in the Way of True Inspiration*. New York: Random House, 2014.

Cavell, Richard. "Gould, McLuhan, and the Fate of the Acoustic." *Glenn Gould Magazine* 10, no. 1 (Spring 2004).

Caves, Richard E. *Creative Industries: Contracts between Art and Commerce*. Cambridge, MA: Harvard University Press, 2000.

Century, Michael. "Cultural Laboratories." In *New Media Culture in Europe*. Amsterdam: Netherlands Design Institute, 1999.

Century, Michael. "Pathways to Innovation in Digital Culture." *Leonardo Electronic Almanac* 11 (November 1999).

Century, Michael, and Thierry Bardini. "Towards a Transformative Set-Up: A Case Study of the Art and Virtual Environments Program at the Banff Centre for the Arts." *Leonardo* 32, no. 4 (1999).

Chapman, Owen, and Kim Sawchuk. "Research-Creation: Intervention, Analysis and 'Family Resemblances.'" *Canadian Journal of Communication* 37, no. 1 (2012).

Charland, Maurice. "Technological Nationalism." *Canadian Journal of Political and Social Theory* 10, no. 1–2 (1986).

Chun, Wendy Hui Kyong. *Programmed Visions: Software and Memory*. Cambridge, MA: MIT Press, 2011.

Churchill, Joan. "An Interview with Evelyn Lambart." *AFISA (Canada)* 15, no. 3 (1987).

Coll, David C. "Yesterday's Adventures with Today's Technology." In *Silicon Valley North: A High-Tech Cluster of Innovation and Entrepreneurship*, edited by Larisa V. Shavinina. Amsterdam: Elsevier, 2004.

Communications Canada. "Communications for the Twenty-First Century: Media and Messages in the Information Age." Ottawa: Communications Canada, 1987.

Cook, Sarah, and Sara Diamond, eds. *Euphoria and Dystopia: The Banff New Media Institute Dialogues*. Banff: Banff Centre Press, Riverside Architectural Press, 2011.

Cox, Geoff. "The Post-Digital and the Problem of Temporality." In *Postdigital Aesthetics: Art, Computation and Design*, edited by David M. Berry and Michael Dieter, 151–162. Basingstoke, UK: Palgrave, 2015.

Crombie, A. C. "Experimental Science and the Rational Artist in Early Modern Europe." *Daedalus: Journal of the American Academy of Arts and Sciences, Art and Science Summer* 116 (Summer 1986).

Cunningham, Merce. *Changes: Notes on Choreography*. New York: Something Else Press, 1968.

Cunningham, Stuart. "Creative Enterprises." In *Creative Industries*, edited by John Hartley. Malden, MA: Blackwell, 2005.

Curtis, David. "Locating McLaren?" *Undercut* 13 (Winter 1984).

Dakroury, Aliaa Ibrahim. "Present at the Creation: The Telecommission Studies and the Intellectual Origins of the Right to Communicate in Canada (1969–71)." PhD diss., Carleton University, 2008.

Darroch, Michael. "The Toronto School: Cross-Border Encounters, Interdisciplinary Entanglements." In *The International History of Communication Study*, edited by Peter Simonsen and David W. Park. London: Routledge, 2015.

Davies, Char. "Letter to Lorne Falk." Banff Centre Archives, 1995-112, 1991.

Davies, Hugh. "Drawn Sound." In *Grove's Dictionary of Music and Musicians*, 2000.

Davis, Todd. "Sat-Tel-Comp: Utopia in Canadian Art and Artist-Run Centres, 1974–1984." Banff Centre Archives, Banff New Media Institute (BNMI) fonds, 2005.

Day, Moira. "The Banff School of Fine Arts 1933–1969: 'A Theater of the People.'" *Alberta Studies in the Arts and Sciences* 2, no. 1 (1989).

Deleuze, Gilles. *Cinema 2: The Time Image*. Translated by Hugh Tomlinson and Robert Galeta. Minneapolis: University of Minnesota, 1986.

Denis, Jean-François. "René Lussier: A Guitarist Who Speaks a Different Language." *Musicworks* 55 (Spring 1994).

Detterer, Gabriele, and Maurizio Nannucci, eds. *Artist-Run Spaces: Nonprofit Collective Organizations in the 1960s and 1970s*. Zurich: JRP Ringier, 2012.

Devon, Terrence. "Interactivity and the Popular Support for Telidon." *Canadian Journal of Communication* 16, no. 2 (1991).

Dhomont, Francis. "But Is This Music?" Paper presented at the >convergence< Canadian Electroacoustic Community, Banff, Alberta, 1989.

Dobson, Terence. "'Bump . . . Bup . . . Bup': Aural Innovation in the Films of Norman McLaren." In *The Music and Sound of Experimental Film*, edited by Holly Rogers and Jeremy Barham. Oxford: Oxford University Press, 2017.

Dobson, Terence. *The Film Work of Norman McLaren*. Eastleigh, UK: John Libbey; Bloomington: Indiana University Press, 2006.

Douglas, Mary. "Rightness of Categories." In *How Classification Works: Nelson Goodman among the Social Sciences*, edited by Mary Douglas and David Hull. Edinburgh: University of Edinburgh Press, 1992.

Downie, Marc. "Choreographing the Extended Agent: Performance Graphics for Dance Theater." PhD diss., Massachusetts Institute of Technology, 2005.

Dufourt, Hugues. "L'artifice d'écriture dans la musique occidentale." *Critique* 408 (1981).

Dumas, Chantal. "Écologie sonore: Le 7e printemps électroacoustique." *Musicworks* 55, no. Spring (1993).

Duplaix, Sophie, ed. *Sons et lumières. Une histoire du son dans l'art du XXe siècle*. Paris: Centre Pompidou, 2004.

Dyson, Frances. *Sounding New Media: Immersion and Embodiment in the Arts and Culture*. Berkeley: University of California Press, 2009.

Eco, Umberto. *The Search for the Perfect Language*. Cambridge, MA: Blackwell, 1995.

Edwards, Paul N. *The Closed World: Computers and the Politics of Discourse in Cold War America*. Cambridge, MA: MIT Press, 1996.

Engelbart, Douglas, and William English. "A Research Centre for Augmenting Human Intellect." Paper presented at the Fall Joint Computer Conference, 1968.

Feenberg, Andrew. *Alternative Modernity*. Berkeley: University of California Press, 1995.

Feinstein, Laura. "Montreal's Flourishing Cyberculture." *Surface*, March 19, 2018. https://www.surfacemag.com/articles/montreal-art-and-technology/.

Fischer, Barbara, and Lorne Falk, eds. *The Event Horizon*. Banff: Walter Phillips Gallery; Toronto: Coachouse Press, 1987.

Fleischer, Alain. "La cinémachine de Michael Snow = Michael Snow's Cinemachine." In *Michael Snow panoramique: Oeuvres photographiques et films*, edited by Piet Coessens et al. Brussels: Société des Expositions du Palais des Beaux-Arts de Bruxelles 1999.

Fleming, Jim, and William Frezza. "NAPLPS: A New Standard for Text and Graphics, Part 1." *Byte*, 1983.

Florida, Richard. *The Rise of the Creative Class: And How It's Transforming Work, Leisure, Community and Everyday Life*. New York: Basic Books, 2002.

Foldes, Peter. "Interview." *Écran*, no. 11 (January 1973): 56–58.

Foley, James D., Andries van Dam, Steven K. Feiner, and John F. Hugues. *Computer Graphics: Principles and Practices*. 2nd ed. Reading, MA: Addison-Wesley, 1987.

Fourmentraux, Jean-Paul. "Governing Artistic Innovation: An Interface among Art, Science and Industry." *Leonardo* 40, no. 5 (2006): 489–492.

Fourmentraux, Jean-Paul. "Linking Art and Sciences, an Organizational Dilemma. About Hexagram Consortium (Montreal, Canada)." *Creative Industries Journal* 3, no. 2 (2010): 137–150.

Fowler, Luke, dir. *Electro-Pythagoras (a Portrait of Martin Bartlett)*. 2017.

Freeman, Christopher. *The Economics of Industrial Innovation*. 3rd ed. London: Pinter, 1997.

Freeman, Christopher. *Long Wave Theory*. Cheltenham, UK: Edward Elgar, 1996.

Freeman, Christopher, and Francisco Louçã. *As Time Goes By: From the Industrial Revolutions to the Information Revolution*. Oxford: Oxford University Press, 2001.

Frenkel, Vera. "Herman and Herbert: The Frye-McLuhan Force-Field." Typescript supplied by the author.

Friis-Hansen, Dana, ed. *Visionary Apparatus: Michael Snow and Juan Geuer*. Cambridge, MA: Hayden Gallery, MIT Committee on the Visual Arts, 1986.

Gagnon, Jean. "Faire comme les musiciens. Le jeu instrumental dans les performances audiovisuelles." PhD diss., Université du Québec à Montréal.

Gagnon, Monika Kin, and Janine Marchessault. *Reimagining Cinema: Film at Expo 67*. Montréal: McGill-Queens University Press, 2015.

Galison, Peter, and Caroline Jones. "Factory, Laboratory, Studio: Dispersing Sites of Production." In *The Architecture of Science*, edited by Peter Galison and Emily Thompson. Cambridge, MA: MIT Press, 1999.

Galloway, Alexander R. *The Interface Effect*. Cambridge, UK: Polity, 2012.

Garro, Diego. "From Sonic Art to Visual Music: Divergences, Convergences, Intersections." *Organised Sound* 17, no. 2 (2012): 103–113.

Gaudreau, Chantal. "Robert Filiou: Une galerie dans une casquette." *Intervention* 6 (1980): 41.

Gell, Alfred. *Art and Agency: An Anthropological Theory*. Oxford: Oxford University Press, 1998.

Gibbons, Michael, et al. *The New Production of Knowledge: The Dynamics of Science and Research in Contemporary Societies*. London: Sage, 1994.

Gillies, James. "Technological Determinism in Canadian Telecommunications: Telidon Technology, Industry and Government." *Canadian Journal of Communication* 15, no. 2 (1990).

Gilpin, Heidi, and Lorne Falk. "Demo Aesthetics." *Convergence* 2 (Autumn 1995): 127–139.

Gleick, James. *The Information: A History, a Theory, a Flood*. New York: Vintage Books, 2012.

Godfrey, David. "Statement on the Canadian Institute for Research on Cultural Enterprise." Banff Centre Archives, 1993-62, Box 2, 1991.

Godin, Benoît, and Dominique Vinck, eds. *Critical Studies of Innovation: Alternative Approaches to the Pro-Innovation Bias*. Cheltenham, UK: Edward Elgar, 2017.

Goehr, Lydia. *The Quest for Voice: On Music, Politics, and the Limits of Philosophy*. Berkeley: University of California Press, 1998.

Goodman, Nelson. "Some Notes on Languages of Art." *Journal of Philosophy* 67, no. 16 (1970): 563–573.

Gopnik, Adam. "The Return of the Native: A Public Intellectual Remakes Himself as a Public Servant." *New Yorker*, September 7, 2009.

Gotlieb, C. C., and Allan Borodin. *Social Issues in Computing*. New York: Academic Press, 1973.

Gould, Glenn. "An Argument for Music in the Electronic Age" (1964). In *The Art of Glenn Gould: Reflections of a Musical Genius*, edited by John P. L. Roberts. Toronto: Malcolm Lester Books, 1999.

Gould, Glenn. "Arnold Schoenberg: A Perspective" (1964). In *The Glenn Gould Reader*, edited by Tim Page. Toronto: Lester & Orpen Dennys, 1984.

Gould, Glenn. "Dialogues on the Prospects of Recording." Canadian Broadcasting Corporation, 1965.

Gould, Glenn. *A Glenn Gould Fantasy*. Compact disc S2K 60686. CBS Records, 1998 (1980).

Gould, Glenn. *Glenn Gould on the Moog Synthesizer*. Canadian Broadcasting Corporation, 1968.

Gould, Glenn. *The Idea of North*. Compact disc PSCD2003-3. CBC Records, 1992 (1967).

Gould, Glenn. "The Prospects of Recording" (1965). In *The Glenn Gould Reader*, edited by Tim Page. Toronto: Lester & Orpen Dennys, 1984.

Gould, Glenn. "Radio as Music: Glenn Gould in Conversation with John Jessop" (1971). In *The Glenn Gould Reader*, edited by Tim Page. Toronto: Lester & Orpen Dennys, 1984.

Gould, Glenn. "Rubinstein" (1965). In *The Glenn Gould Reader*, edited by Tim Page. Toronto: Lester & Orpen Dennys, 1984.

Gould, Glenn. "Where Music and Film Meet—Glenn Gould in Conversation with Norman McLaren." *Glenn Gould Magazine* 8, no. 1 (Spring 2002).

Government of Canada Department of Communications and National Film Board. *Telidon*. Toronto: Lauron Productions, 1981.

Graham, Gerald. *Canadian Film Technology: 1896–1986*. Newark: University of Delaware Press, 1989.

Grau, Oliver. *Virtual Art: From Illusion to Immersion*. Rev. and expanded ed. Cambridge, MA: MIT Press, 2003.

Greelish, David. "An Interview with Computing Pioneer Alan Kay." *Time*, April 2, 2013.

Guédon, Jean-Claude. "Norme ou système technique? Les avatars de Télidon au Canada." *Réseaux* 7, no. 37 (1989).

Halpern, Orit. *Beautiful Data: A History of Vision and Reason since 1945*. Durham: Duke University Press, 2014.

Hankins, Thomas, and Robert J. Silverman. *Instruments and the Imagination*. Princeton: Princeton University Press, 1995.

Hansen, Mark B. N. *Bodies in Code: Interfaces with Digital Media*. New York: Routledge, 2006.

Hansen, Mark B. N. *New Philosophy for New Media*. Cambridge, MA: MIT Press, 2004.

Hébert, Pierre. *L'ange et l'automate. Propos sur le cinéma d'animation et autres sujets.* Laval, PQ: Les 400 coups, 1999.

Hébert, Pierre. "Considérations très personnelles sur l'état du cinéma d'animation." *24 Images* 106 (2001).

Hébert, Pierre. "Les enjeux de l'art à l'ère des machines. Effacement et résistance du corps." *Cinéma 24 Images. La revue québecoise du cinéma,* Summer 1989.

Heble, Ajay. "New Contexts of Canadian Criticism: Democracy, Counterpoint, Responsibility." In *New Contexts of Canadian Criticism,* edited by Ajay Heble, Donna Palmateer Pennee, and J. R. Tim Struthers, 85–93. Peterborough, ON, and Guelph, ON: Broadview Press, 1997.

Heim, Michael. *Virtual Realism.* New York: Oxford, 1998.

Helmuth, Mara. "Barry Truax's *Riverrun.*" In *Analytical Methods of Electroacoustic Music,* edited by Mary Simoni. New York: Routledge, 2006.

Henighan, Tom. *The Presumption of Culture: Structure, Strategy, and Survival in the Canadian Culture Landscape.* Vancouver: Raincoast Books, 1996.

Higgins, Hannah, and Douglas Kahn. *Mainframe Experimentalism: Early Computing and the Foundations of the Digital Arts.* Berkeley: University of California Press, 2012.

Hill, Richard. "Photo/Electric Art Show." In *Informatique/Culture, Computer Culture 83.* Villeneuve-lez-Avignon, France: La Chartreuse/Centre International de Recherche de Création et d'Animation, 1983.

Hiltzik, Michael. *Dealers of Lightning: Xerox Parc and the Dawn of the Computer Age.* New York: HarperCollins, 1999.

Hobsbawm, Eric. *The Age of Extremes: The Short Twentieth Century, 1914–1989.* London: Michael Joseph, 1994.

Holmes, Thom, and Terence M. Pender. *Electronic and Experimental Music: Technology, Music, and Culture.* New York: Routledge, 2012.

Howes, David. "Glenn Gould's Contrapuntal Constitution." http://canadianicon.org/table-of-contents/glenn-goulds-contrapuntal-constitution/.

Hughes, Thomas Parke. *Rescuing Prometheus.* New York: Pantheon Books, 1998.

Huhtamo, Erkki. "Time Traveling in the Gallery: An Archeological Approach to Media Art." In *Immersed in Technology: Art and Virtual Environments,* edited by Mary Anne Moser and Douglas MacLeod. Cambridge, MA: MIT Press, 1996.

Hui, Yuk, and Andreas Broeckmann. "Introduction." In *30 Years after Les Immatériaux: Art, Science and Theory,* edited by Yuk Hui and Andreas Broeckmann. Lüneburg: Leuphana University Lüneburg, Centre for Digital Cultures, 2014.

Isaacson, Walter. *The Innovators: How a Group of Inventors, Hackers, Geniuses, and Geeks Created the Digital Revolution*. New York Simon & Schuster, 2014.

Jarvis, Douglas. "Sat-Tel-Comp: The Cost of Humanizing Technology." Banff Centre Archives, Banff New Media Institute (BNMI) fonds: Paul D. Fleck Library and Archives at Banff Centre for Arts and Creativity, 2005.

Jha, Alok. "Computer-Generated Realities Are Becoming Ubiquitous." *Economist*, October 20, 2020.

Johnston, Andrew R. "Models of Code and the Digital Architecture of Time." *Discourse* 37, no. 3 (Fall 2015): 221–246.

Johnston, Hugh. *Radical Campus: Making Simon Fraser University*. Vancouver: D & M Publishers, 2009.

Jones, Caroline, and Peter Galison. *Picturing Science, Producing Art*. New York: Routledge, 1998.

Jones-Imhotep, Edward. "Malleability and Machines: Glenn Gould and the Technological Self." *Technology and Culture* 57, no. 2 (April 2016): 287–321.

Joseph, Branden. *Beyond the Dream Syndicate: Tony Conrad and the Arts after Cage*. New York: Zone Books, 2008.

Kac, Eduardo. *Telepresence and Bio Art: Networking Humans, Rabbits and Robots*. Ann Arbor: University of Michigan Press, 2005.

Kaganskiy, Julia. "Our Sensorium: New Ways of Seeing and Perceiving." *Neural* 59 (2018).

Kahn, Douglas. "Sound Art, Art, Music." *Iowa Review Web* 7, no. 1 (2005).

Kane, Carolyn L. *Chromatic Algorithms: Synthetic Color, Computer Art, and Aesthetics after Code*. Chicago: University of Chicago Press, 2014.

Kay, Alan. "The Early History of Smalltalk." In *History of Programming Languages II*, edited by Thomas J. Bergin and Richard G. Gibson. New York: ACM Press; Reading, MA: Addison-Wesley, 1996.

Kay, Alan. "User Interface: A Personal View." In *The Art of Human-Computer Interface Design*, edited by Brenda Laurel. Reading, MA: Addison-Wesley, 1990.

Kay, Alan, and Adele Goldberg. "Personal Dynamic Media." *IEEE Computer* 10 (March 1977): 31–41.

Keck, Jennifer Marguerite. "Making Work: Federal Job Creation Policy in the 1970s." PhD diss., University of Toronto, 1995.

Kelman, Ari Y. "Rethinking the Soundscape. A Critical Genealogy of a Key Term in Sound Studies." *Senses and Society* 5, no. 2 (2010): 212–234.

Khonsary, Jeff, and Kristina Lee Podesva, eds. *Institutions by Artists*. Vol. 1. Vancouver: Fillip Editions, Folio C, 2012.

King, Sarah. "In/Consequential Relationships: Refusing Colonial Ethics of Engagement in Yuxweluptun's Inherent Rights, Vision Rights." *British Columbia Studies* 10, no. 193 (2017).

Kingwell, Mark. *Glenn Gould*. Toronto: Penguin Canada, 2009.

Kittler, Friedrich. *Gramophone, Film, Typewriter*. Translated by Geoff Winthrop-Young and Michael Wutz. Stanford: Stanford University Press, 1997.

Kluitenberg, Eric, ed. *Book of Imaginary Media: Excavating the Dream of the Ultimate Communication Medium*. Amsterdam: NAi Publishers, 2006.

Klüver, Billy, Julie Martin, and Barbara Rose. *Pavilion: Experiments in Art and Technology*. New York: Dutton, 1972.

Knorr-Cetina, Karin. *Epistemic Cultures: How the Sciences Make Knowledge*. Cambridge, MA: Harvard University Press, 1999.

Kochanek, Doris. "A Computer System for Smooth Keyframe Animation." M.Sc., University of Waterloo, 1982.

Kroker, Arthur. *Technology and the Canadian Mind: Innis/McLuhan/Grant*. Montréal: New World Perspectives, 1984.

Kuffert, L. B. *A Great Duty: Canadian Responses to Modern Life and Mass Culture, 1939–1967*. Montréal: McGill-Queen's University Press, 2003.

Kwastek, Katja. *Aesthetics of Interaction in Digital Art*. Cambridge, MA: MIT Press, 2013.

Lander, Dan, and Micah Lexier, eds. *Sound by Artists*. Toronto: Art Metropole; Banff: Walter Phillips Gallery, 1990.

Langill, Caroline. "Shifting Polarities: Exemplary Works of Canadian Electronic Media Art Produced between 1970 and 1991." Fondation Daniel Langlois, http://www.fondation-langlois.org/html/e/page.php?NumPage=1949.

Langill, Caroline Seck. "The Telidon Trials (1978–1985): Creative Applications and Disappointing Outcomes (for Some)." Paper presented at Rewire, the Fourth International Conference on the Histories of Media Art, Science and Technology, Liverpool, 2011.

Langill, Caroline. "Vera Frenkel's *String Games* and the Germination of Media Art in Canada." In *Esthétique des arts médiatiques. Ensemble ailleurs/Together Elsewhere*, edited by Louise Poissant and Pierre Tremblay. Québec: Presses de l'Université du Québec, 2010.

Lanier, Jaron. *Dawn of the New Everything*. New York: Random House, 2017.

Latour, Bruno. "Visualisation and Cognition: Thinking with Eyes and Hands." *Knowledge and Society* 6 (1986): 1–40.

Laurel, Brenda. "Placeholder: Notes on Process." http://www.tauzero.com/Brenda_Laurel/Severed_Heads/Placeholder_Process_Notes.html.

Laurel, Brenda, ed. *The Art of Human-Computer Interface Design*. Reading, MA: Addison-Wesley, 1990.

Lee, S. K., W. Buxton, and K. C. Smith. "A Multi-Touch Three Dimensional Touch-Sensitive Tablet." Paper presented at the Proceedings of the ACM Conference on Human Factors in Computing Systems (CHI'85), 1985.

Lefebvre, Marie-Thérèse. "L'underground musical des années 1970 au Québec: Émergence des pratiques en 'musique actuelle' et rencontre entre deux univers de création." *Cahiers des dix*, no. 68 (2014): 249–286.

Leler, William. "Actor-Based Simulation + Linda = Virtual Environments." Paper presented at the Second Conference on Cyberspace, Santa Cruz, CA, 1991.

Lenoir, Timothy. "Virtual Reality Comes of Age." In National Research Council, *Funding a Revolution: Government Sponsorship for Computing Research*. Washington, DC: National Academies Press, 1999.

Lepore, Jill. "The Disruption Machine: What the Gospel of Innovation Gets Wrong." *New Yorker*, June 23, 2014.

Lévi-Strauss, Claude. *The Savage Mind*. Chicago: University of Chicago Press, 1966.

Lewin-Richter, Andrés, and Alcides Lanza. "On Hugh Le Caine." *Musicworks* 83 (Summer 2002).

Lewis, George E. "Improvising Tomorrow's Bodies: The Politics of Transduction." Hemispheric Institute. http://hemi.nyu.edu/journal/4.2/eng/en42_pg_lewis.html.

Lewis, George E. "Noise on the Western Front: Music and Sound around Mount Pleasant." Vancouver. https://front.bc.ca/events/noise-on-the-western-front-music-and-sound-around-mount-pleasant/.

Lewis, George E. "Real Time Composing and Improvising Machines." In *Viewpoints on Art and Technology*. Banff Centre Archives File IA117b. Banff, 1986.

Lewis, George E. "Toni Dove's Nonmodern Ontologies." In *Toni Dove: Embodied Machines*. Sarasota, FL: Ringling Museum of Art, 2018.

Lewis, George E. "Too Many Notes: Computers, Complexity, and Culture in Voyageur." *Leonardo Music Journal* 10 (2000): 33–39.

Lewis, Glenn. "The Living Museums Data Network." In "Spaces by Artists," edited by Tanya Rosenberg, *Parallelogram* 3 (1979): 118–136.

Licklider, J. C. R. "Man-Machine Symbiosis." *IRE Transactions on Human Factors in Electronics* (March 1960). Reprinted in *The New Media Reader*, edited by Noah Wardrip-Fruin and Nick Montfort (Cambridge, MA: MIT Press, 2003).

Licklider, J. C. R. "A Picture Is Worth a Thousand Words—and It Costs . . ." Paper presented at the Spring Joint Computer Conference, Boston, 1969.

Locke, John W. "Michael Snow's 'La Région Centrale': How You Should Watch the Best Film I Ever Saw." *Artforum* 12, no. 4 (1973).

Low, Joni. "Connections/Disconnections." In *Hank Bull: Connexion*, edited by Pan Wendt. Charlottetown, PE: Confederation Centre Art Gallery, Burnaby Art Gallery, 2015.

Lowry, Glen. "Props to Bad Artists: On Research-Creation and a Cultural Politics of University-Based Art." *Canadian Art Review (Revue d'art canadienne)* 40, no. 1 (2015).

Loy, Gareth. "Musicians Make a Standard: The MIDI Phenomenon." *Computer Music Journal* 9, no. 4 (1985): 8–26.

Lozano-Hemmer, Rafael. "Perverting Technological Correctness." *Leonardo* 29, no. 1 (1996): 5–15.

Lundvall, Bengt-Åke. "National Innovation Systems—Analytical Concept and Development Tool." *Industry and Innovation* 14 (2007): 95–119.

Lunenfeld, Peter. "Cyborg Economics: Demo or Die." In *Snap to Grid: A User's Guide to Digital Arts, Media, and Cultures*. Cambridge, MA: MIT Press, 2000.

Lyman, Peter. *Canada's Video Revolution: Pay-TV, Home Video and Beyond*. Toronto: James Lorimer, 1983.

Manning, Erin. *Politics of Touch: Sense, Movement, Sovereignty*. Minneapolis: University of Minnesota Press, 2007.

Manovich, Lev. *The Language of New Media*. Cambridge, MA: MIT Press, 2001.

Mantere, Juha Markus. *The Gould Variations: Technology, Philosophy and Criticism in Glenn Gould's Musical Thought and Practice*. Frankfurt: Peter Lang, 2012.

Martin, André. "Cinéma d'animation et ordinateur." *Cinéma/Québec*, 1976.

Martin, André. "Le mystère d'un cinéma instrumental." *Cahiers du cinéma* 81–83 (February-April 1958).

Martin, Ben R. "The Evolution of Science Policy and Innovation Studies." *Research Policy* 41, no. 7 (2012): 1219–1239.

McKinnie, Michael. "Bees, Horseshoes, and Puppets for the Elderly: The Local Initiatives Program and the Political Economy of Canadian Theatre." *Contemporary Theatre Review* 15, no. 4 (2005): 427–439.

McLaren, Norman. "Animated Films." *Documentary Film News* (1948): 52–53.

McLuhan, Marshall. "Canada as Counter-Environment" (1989). In *Canadian Cultural Studies: A Reader*, edited by Sourayan Mookerjea, Imre Szeman, and Gail Faurschou. Durham: Duke University Press, 2009.

McLuhan, Marshall. "Canada: The Borderline Case." In *The Canadian Imagination: Dimensions of a Literary Culture*, edited by David Staines. Cambridge, MA: Harvard University Press, 1977.

McLuhan, Marshall. "Media as Art Forms." *Explorations* 2 (April 1954): 6–13.

McLuhan, Marshall. *Understanding Media, the Extensions of Man*. New York: McGraw Hill, 1964.

McLuhan, Marshall. *Letters of Marshall McLuhan*. Edited by Matie Molinaro, Corrine McLuhan, and William Toye. Toronto: Oxford University Press, 1987.

McWilliams, Donald. *Creative Process: Norman McLaren*. Montréal: National Film Board of Canada, 1990.

McWilliams, Donald. "A God in Volda." *Journal of Film Preservation*, no. 69 (2005): 29–39.

McWilliams, Donald, dir. *On the Creative Process*. Montréal: National Film Board of Canada, 1991.

Mezei, Leslie. "Computer Art." *ArtsCanada*, 1968.

Mezei, Leslie, and Arnold Rockman. "The Electronic Computer as an Artist." *Arts Canada*, no. 94 (November-December 1964): 365–367.

Milz, Sabine. "Canadian Cultural Policy-Making at a Time of Neoliberal Globalization." *ESC: English Studies in Canada* 33, no. 1 (2007): 85–107.

Mintzberg, Henry, and Alexandra McHugh. "Strategy Formation in an Adhocracy." McGill University, 1983.

Mitchell, William J., Alan S. Inouye, and Marjorie S. Blumenthal, eds. *Beyond Productivity: Information Technology, Innovation, and Creativity*. Washington, DC: National Academies Press, 2003.

Mollaghan, Aimée. "An Experiment in Pure Design: The Minimalist Aesthetic in the Line Films of Norman McLaren." *Animation Studies* 6 (2011).

Monahan, Gordon. "A Half Hour Behind Means You're Slightly Ahead. A Report on the Newfoundland Sound Symposium." *Musicworks* 36 (1986).

Moores, Margaret. *Art Creates Change*. V/Tape (Distributor), 1983.

Moretti, Pierre. "Memorandum to René Jodoin." File # 73 009. National Film Board of Canada Archives, 1971.

Mosco, Vincent. *Pushbutton Fantasies: Critical Perspectives on Videotex and Information Technology*. Communication and Information Science. Norwood, NJ: Ablex, 1982.

Moser, Mary Anne, and Douglas MacLeod, eds. *Immersed in Technology: Art and Virtual Environments*. Cambridge, MA: MIT Press, 1996.

Moser, Mary Anne, Catherine Richards, and Nell Tenhaaf, eds. *Virtual Seminar on the Bioapparatus*. Banff: Banff Centre for the Arts, 1991.

Munster, Anna. *Materializing New Media: Embodiment in Information Aesthetics*. Hanover, NH: Dartmouth College Press, 2006.

Nattiez, Jean-Jacques. "Gould Out of Time." Translated by Denis D. Coupal. In *Glenn Gould: Universe of a Genius*, edited by Ghyslaine Guertin. Boisbriand, PQ: Momentum, 2007.

Nelson, Ted. *Computer Lib/Dream Machines*. Chicago: Nelson, 1974.

Nemiroff, Diana. "A History of Artist-Run Spaces in Canada, with Particular Reference to Véhicule, A Space and the Western Front." Master's thesis, Concordia University, 1985.

Nemiroff, Diana. "Par-Al-Lel." *Parallelogram* 9, no. 1 (1983): 16–19.

Nemiroff, Diana, Robert Houle, and Charlotte Townsend-Gault. *Land, Spirit, Power: First Nations at the National Gallery of Canada*. Ottawa: National Gallery of Canada, 1992.

Normandeau, Robert. ". . . Et vers un cinéma pour l'oreille." *Circuit: Musiques Contemporaines* 4, no. 1–2 (1993).

Nowotny, Helga. "Beyond Innovation. Temporalities. Re-Use. Emergence." Gifford Lecture, University of Edinburgh, 2015. https://www.giffordlectures.org/lectures/beyond-innovation-temporalities-re-use-emergence.

Nowotny, Helga. *Insatiable Curiosity: Innovation in a Fragile Future*. Translated by Mitch Cohen. Cambridge, MA: MIT Press, 2008.

Okrent, Arika. *In the Land of Invented Languages: Esperanto Rock Stars, Klingon Poets, Loglan Lovers, and the Mad Dreamers Who Tried to Build a Perfect Language*. New York: Spiegel & Grau, 2009.

Oliveros, Pauline. "Composing with Hugh Le Caine's Special Purpose Tape Recorder at UTEMS in 1966." *eContact* 17, no. 3 (2015).

Oliveros, Pauline. *Sonic Meditations*. N.p.: Smith Publications/American Music, 1974.

Orrghen, Anna. "Surveying the Literature on Technoscience Art: From Pioneer Stories to Collaborations between Artists, Scientists and Engineers as the Object of Study." *Digital Creativity* 28, no. 2 (2017): 157–176.

Ostry, Bernard. *The Cultural Connection: An Essay on Culture and Government Policy in Canada*. Toronto: McClelland and Stewart, 1978.

Ostwald, Peter. *Glenn Gould: The Ecstasy and Tragedy of Genius*. New York: W. W. Norton, 1997.

Oswald, John. "Bad Relations: Plunderography, Pop and Weird in DAB." Paper presented at the >convergence< Canadian Electroacoustic Community, Banff, 1989.

Oswald, John. "Plunderphonics: Or, Audio Piracy as a Compositional Prerogative." *Musicworks* 34 (Spring 1986).

Our World. "Our World Television Broadcast June 25, 1967." http://www.cbc.ca/archives/entry/our-world-five-continents-linked-via-satellite.

Panofsky, Erwin. "Artist, Scientist, Genius: Notes on the 'Renaissance-Dämmerung.'" In *The Renaissance: A Symposium*. New York: Metropolitan Museum of Art, 1952.

Parikka, Jussi. "The Lab Imaginary: Speculative Practices in Situ." In *Across and Beyond*, edited by Kristoffer Gansing, Ryan Bishop, Jussi Parrika, and Elvia Wilk, 76–89. Berlin: Sternberg, 2017.

Parke, Frederic. "Control Parameterization for Facial Animation." In *Computer Animation*. Vienna: Springer-Verlag, 1991.

Patterson, Zabet. "From the Gun Controller to the Mandala: The Cybernetic Cinema of John and James Whitney." *Grey Room*, no. 36 (Summer 2009).

Payzant, Geoffrey. *Glenn Gould: Music and Mind*. Toronto: Van Nostrand Reinhold, 1992.

Penny, Simon. "Improvisation and Interaction, Canons and Rules, Emergence and Play." In *The Oxford Handbook of Critical Improvisation Studies*, edited by George Lewis and Benjamin Piekut, 2 vols., 2:401–423. New York: Oxford University Press, 2016.

Perez, Carlota. "Technological Revolutions and Techno-Economic Paradigms." *Cambridge Journal of Economics* 34, no. 1 (2010).

Perry, Bill. "Computerese." In *Art Is Communications*, edited by Paul Petro. Toronto: United Media Art Studies (Toronto Community Videotex), 1985.

Piché, Jean. "Music on Image: Compositional Strategies for a New Art Form." *Journal of the Acoustical Society of America* 104, no. 1780 (1998).

Pickering, Andrew. *The Cybernetic Brain: Sketches of Another Future*. Chicago: University of Chicago Press, 2010.

Pickering, Andrew. "The Mangle of Practice: Agency and Emergence in the Sociology of Science." In *The Science Studies Reader*, edited by Mario Biagioli, 372–393. New York: Routledge, 1999.

Poggioli, Renato. *The Theory of the Avant-Garde*. Translated by Gerald Fitzgerald. Cambridge, MA: Harvard University Press, 1968.

Powe, B. W. *A Canada of Light*. Toronto: Somerville, 1993.

Powe, B. W. *Mystic Trudeau*. Toronto: Thomas Allen, 2007.

Prévost, Hélène. "Écoutes urbaines." In *S:On. Sound in Contemporary Canadian Art*, edited by Nicole Gingras. Montréal: Artexte Éditions, 2003.

Price, David A. *The Pixar Touch: The Making of a Company*. New York: Knopf, 2008.

Pulfer, J. K. "Man-Machine Communication in Creative Applications." *International Journal of Man-Machine Studies* 3 (1971): 1–11.

Purcell, Patrick. "Digital Media and the Primacy of Content." *IEEE Multimedia* (1997).

Raboy, Marc. "Communication and Globalization: A Challenge for Public Policy." In *Street Protests and Fantasy Parks: Globalization, Culture, and the State*, edited by David R. Cameron and Janice Gross Stein. Vancouver: University of British Columbia Press, 2002.

Raboy, Marc. "Pas de deux: Media Policy and Cultural Politics in Canada and Quebec." In *Media Policy, National Identity and Citizenry in Changing Democratic Societies: The Case of Canada*, edited by Joel Smith, 90–103. Durham: Canadian Studies Center, Duke University, 1998.

Raboy, Marc, and Jeremy Shtern. *Media Divides: Communication Rights and the Right to Communicate in Canada*. Vancouver: University of British Columbia Press, 2010.

Radiolab. "Mr. Bliss." Edited by Tom Howard. WNYC New York, 2012.

Randolph, Jeanne. "A City for Bachelors." In *Immersed in Technology: Art and Virtual Environments*, edited by Mary Anne Moser and Douglas MacLeod. Cambridge, MA: MIT Press, 1996.

Randolph, Jeanne. "Influencing Machines: The Relationship between Art and Technology." In *Psychoanalysis and Synchronized Swimming, and Other Writings on Art*. Toronto: YYZ Books, 1991.

Reichardt, Jasia. *Cybernetics, Art, and Ideas*. Greenwich, CT: New York Graphic Society, 1971.

Riar, Inderbir Singh. "Expo 67, or the Architecture of Late Modernity." PhD diss., Columbia University, 2014.

Richards, Catherine. "Body in Ruins." In *Body Invaders: Panic Sex in America*, edited by Arthur Kroker and Marilouise Kroker. New York: St. Martin's Press, 1987.

Richards, Catherine. "Virtual Worlds, Digital Images." In *1987 AFI Video Festival*, edited by Kenneth Kirby, 66–67. Los Angeles: American Film Institute, 1987.

Robertson, Clive. *Policy Matters: Administrations of Art and Culture*. Toronto: YYZ Books, 2006.

Robineau, Anne. "La musique 'actuelle' québécoise. Une façon distincte de développer une pratique artistique." *Études canadiennes/Canadian Studies* 66 (2009): 105–113.

Robinett, Warren, and Michael Naimark. "Artists Explore Virtual Reality: The Bioapparatus Residency at the Banff Centre for the Arts." *Presence: Teleoperators and Virtual Environments* 1, no. 2 (1992): 248–250.

Rokeby, David. "Context-Awareness and Meaning—an Interview with David Rokeby." By Ulrik Ekman (2014). http://ubiquity.dk/?page_id=175.

Rokeby, David. "The Harmonics of Interaction." *Musicworks* 46 (Spring 1990).

Rokeby, David. "Transforming Mirrors: Subjectivity and Control in Interactive Media." In *Critical Issues in Electronic Media*, edited by Simon Penny. Albany: State University of New York Press, 1995.

Rosebush, Judson. "The Proceduralist Manifesto." *Leonardo* 2, SIGGRAPH special supplement (1989).

Rothman, Joshua. "Are We Already Living in Virtual Reality?" *New Yorker*, April 2, 2018.

Rosenboom, David. "On Being Invisible." *Musicworks* 28 (Summer 1984).

Rosenboom, David, ed. *Biofeedback and the Arts: Results of Early Experiments* Vancouver: Aesthetic Research Centre of Canada, 1976.

Rosenboom, David. *Invisible Gold*. CD POG022. Pogus Productions, 2000.

Rosenthal, Alan. "Norman McLaren on Pas de Deux." *Journal of the University Film Association* 122, no. 1 (1970): 15–20.

Rotman, Brian. "Corporeal or Gesturo-Haptic Writing." *Configurations* 10, no. 2 (2002): 423–438.

Rutherdale, Myra, and Jim Miller. "'It's Our Country': First Nations' Participation in the Indian Pavilion at Expo 67." *Journal of the Canadian Historical Association / Revue de la Société historique du Canada* 17, no. 2 (2006).

Ryle, Gilbert. "Improvisation." *Mind* 85, no. 337 (1976).

Said, Edward. "On Glenn Gould: Remembrances of Things Played: Presence and Memory in the Pianist's Art." In *Reflections on Exile and Other Essays*. Cambridge, MA: Harvard University Press, 2002.

Salter, Chris. *Entangled: Technology and the Transformation of Performance*. Cambridge, MA: MIT Press, 2010.

Salter, Chris, Regula Valérie Burri, and Joseph Dumit. "Art, Design, and Performance." In *The Handbook of Science and Technology Studies*, 4th ed., edited by Rayvon Fouché, Ulrike Felt, Clark A. Miller, and Laurel Smith-Doerr. Cambridge, MA: MIT Press, 2016.

Saunders, Doug. "Canada 150: In 1967, Change in Canada Could No Longer Be Stopped." *Globe and Mail*, January 6, 2017.

Sawchuk, Kim. "An Index of Power: Innis, Aesthetics and Technology." In *Harold Innis in the New Century: Reflections and Refractions*, edited by William J. Buxton and Charles R. Acland. Montréal: McGill-Queen's University Press, 1999.

Sawchuk, Kim. "Media Arts Revisited." *Canadian Journal of Communication* 37, no. 4 (2011).

Sawchuk, W., H. G. Bown, D. O'Brien, and G. W. Thorgeirson. "An Interactive Image Communication System Using Narrowband Lines." *Computers and Graphics* 3 (1978).

Schafer, R. Murray. "Acoustic Design." *Musicworks* 25 (1983).

Schafer, R. Murray. *On Canadian Music*. Bancroft, ON: Arcana Editions, 1984.

Schafer, R. Murray. *String Quartet 2*. Toronto: Berandol, 1976.

Schafer, R. Murray. *The Tuning of the World*. Toronto: McClelland and Stewart, 1977.

Schafer, R. Murray. *Voices of Tyranny, Temples of Silence*. Indian River, ON: Arcana Editions, 1993.

Scheidt, Dan. "Computed Music." *Musicworks* 28 (Summer 1984).

Schiphorst, Thecla. "A Case Study of Merce Cunningham's Use of the Lifeforms Computer Choreographic System in the Making of Trackers." Master's thesis, Simon Fraser University, 1993.

Schlesinger, Philip. "The Creative Economy: Invention of a Global Orthodoxy." *Innovation: The European Journal of Social Science Research* 30, no. 1 (2017): 73–90.

Schön, Donald A. *The Reflective Practitioner: How Professionals Think in Action*. New York: Basic Books, 1983.

Schryer, Claude. "Electroacoustic Sound Ecology: A Report on the Seventh Printemps Électroacoustique." *Musicworks* 55 (Spring 1993).

Shaw, Nancy. "Expanded Consciousness and Company Types: Collaboration since Intermedia and N.E. Thing Co." In *Vancouver Anthology: The Institutional Politics of Art*, edited by Stan Douglas. Vancouver: Talon Books, 1991.

Shea, Geoffrey. "Dead Medium: Canada's Telidon Network." http://www.deadmedia.org/notes/2/024.html.

Sherman, Tom. "Cultural Engineering." Ottawa: SAW Gallery, 2016.

Shulgin, Alexei. "Art, Power, and Communication." https://v2.nl/archive/articles/art-power-and-communication.

Sifianos, Georges. "The Definition of Animation: A Letter from Norman McLaren." *Animation Journal* 3, no. 2 (Spring 1995).

Simondon, Gilbert. "Gilbert Simondon entretien sur la mécanologie avec Jean Le Moyne" (1968). *Revue de synthèse*, 6th ser., 130, no. 1 (2009): 103–132.

Simondon, Gilbert. *On the Mode of Existence of Technical Objects*. Translated by Ninian Mellamphy. London, ON: University of Western Ontario, 1980 (1958).

Sitney, P. Adams. *Visionary Film: The American Avant-Garde, 1943–2000*. 3rd ed. Oxford: Oxford University Press, 2002.

Smith, Alvy Ray. "The Stuff of Dreams (25 Years = 100,000x)." *Computer Graphics World* (July 1998): 27–29.

Smith, Chris. *Creative Britain*. London: Faber and Faber, 1998.

Smith, David Canfield. "Pygmalion: A Creative Programming Environment." PhD diss., Stanford University, 1975.

Snow, C. P. *The Two Cultures and the Scientific Revolution*. Cambridge: Cambridge University Press, 1959.

Snow, Michael. *The Collected Writings of Michael Snow*. Waterloo, ON: Wilfred Laurier University Press, 1994.

Star, Susan Leigh. "This Is Not a Boundary Object: Reflections on the Origin of a Concept." *Science, Technology, and Human Values* 35, no. 5 (2010): 601–617.

Stark, David. *The Sense of Dissonance: Accounts of Worth in Economic Life*. Princeton: Princeton University Press, 2011.

Steinmueller, W. Edward. "The U.S. Software Industry: An Analysis and Interpretive History." In *The International Computer Software Industry*, edited by David C. Mowery. Oxford: Oxford University Press, 1996.

Sterling, C. H. "Pioneering Risk: Lessons from the US Teletext/Videotex Failure." *Annals of the History of Computing* 28, no. 3 (2006).

Sterne, Jonathan. *The Audible Past: Cultural Origins of Sound Reproduction*. Durham: Duke University Press, 2003.

Sterne, Jonathan. "Compression: A Loose History." In *Signal Traffic: Critical Studies of Media Infrastructures*, edited by Lisa Parks and Nicole Starosielski. Urbana: University of Illinois Press, 2015.

Sterne, Jonathan. *MP3: The Meaning of a Format*. Durham: Duke University Press, 2012.

Sterne, Jonathan. "Sonic Imaginations." In *The Sound Studies Reader*. New York: Routledge, 2012.

Suchman, Lucy. "The End of Innovation (as We Knew It): A Contribution to Critical Innovation Studies." University of California Research Institute, http://rewired.uchri.org/index.html%3Fp=109.html.

Suchman, Lucy. *Human-Machine Reconfigurations: Plans and Situated Actions*. Cambridge: Cambridge University Press, 2017.

Suchman, Lucy, and Libby Bishop. "Problematizing 'Innovation' as a Critical Project." *Technology Analysis and Strategic Management* 12, no. 3 (2000): 327–333.

Sutherland, Ivan E. "Sketchpad: A Man-Machine Graphical Communication System." MIT, 1963.

Sutton, Gloria. "Stan VanDerBeek's *Poemfields*: The Interstice between Cinema and Computing." In *Mainframe Experimentalism: Early Computing and the Foundations of the Digital Arts*, edited by Hannah B. Higgins and Douglas Kahn, 311–333. Berkeley: University of California Press, 2012.

Tal, Joseph. "An Encounter with Hugh Le Caine." *Musicworks* 83 (Summer 2002).

Taylor, Charles. "Can Canada Survive the Charter?" *Alberta Law Review* 30 (1992): 427–447.

Taylor, Charles. "Deep Diversity and the Future of Canada." In *Can Canada Survive? Under What Terms and Conditions?*, edited by David M. Hayne. Toronto: University of Toronto Press, 1997.

Taylor, James R. "The Twentieth Century in the Rear View Mirror." *Canadian Journal of Communications* 13, no. 3/4 (1988).

Telecommission. *Instant World (Sans Frontières): A Report on Telecommunications in Canada*. Ottawa: Information Canada, 1971.

Testa, Bart. "An Axiomatic Cinema: Michael Snow's Films." In *The Michael Snow Project: Presence and Absence—the Films of Michael Snow 1956–1991*, edited by Jim Shedden. Toronto: Art Gallery of Ontario, 1995.

Thain, Allana. "In the Blink of an Eye: Norman McLaren between Dance and Animation." In *The Oxford Handbook of Screendance Studies*, edited by Douglas Rosenberg. Oxford: Oxford University Press, 2016.

Theall, Donald. "Expo 67 as Total Environment." In *Reimagining Cinema: Film at Expo 67*, edited by Monika Kin Gagnon and Janine Marchessault. Montréal: McGill-Queens University Press, 2015.

Théberge, Paul. *Any Sound You Can Imagine: Making Music/Consuming Technology*. Hanover, NH: University Press of New England, 1997.

Thomas, Frank. *Disney Animation: The Illusion of Life*. New York: Abbeville Press, 1981.

Thompson, Gordon B. "Computers and Visual Literacy." Paper presented at the Third Man-Computer Communication Seminar, Ottawa, 1973.

Thompson, Gordon B. "Visual Literacy and the Time-Varying Icon." *IEEE Journal on Selected Areas in Communications* 1, no. 2 (1983).

Thibault, Ghislain, and Mark Hayward. "Jean Le Moyne's *Itinéraire mécanologique*: Machine Poetics, Reverie, and Technological Humanism." *Canadian Literature* 221 (Summer 2014): 56–72.

Tomas, David. "Reflections beyond the Interface of Theory and Practice." In *Virtual Seminar on the Bioapparatus*, edited by Mary Anne Moser, Catherine Richards, and Nell Tenhaaf. Banff: Banff Centre for the Arts, 1991.

Tomas, David. *Vertov, Snow, Farocki: Machine Vision and the Posthuman*. New York: Bloomsbury Academic, 2013.

Truax, Barry. *Acoustic Communication*. 2nd ed. Westport, CT: Ablex, 2001.

Truax, Barry. "Computer Music and Acoustic Communication: Two Emerging Interdisciplines." In *Outside the Lines: Issues in Interdisciplinary Research*, edited by Liora Salter and Alison Hearn. Montréal: McGill-Queen's University Press, 1996.

Truax, Barry. "Editorial: Context-Based Composition." *Organised Sound* 22, no. 1 (2017): 1–3.

Truax, Barry, ed. *Handbook for Acoustic Ecology*. Vancouver: Aesthetic Research Centre, 1978.

Truax, Barry. "Interview by Michael E. Karman." 2009. Formerly available at http://asymmetrymusicmagazine.com/interviews/barry-truax/.

Truax, Barry. "Real-Time Granular Synthesis with a Digital Signal Processor." *Computer Music Journal* 12, no. 2 (1988): 14–26.

Truax, Barry. "Soundscape, Acoustic Communication and Environmental Sound Composition." *Contemporary Music Review* 15, no. 1–2 (1996).

Tuer, Dot. "Beyond the New Media Frame: The Poetics of Absence in Vera Frenkel's String Games." In *Vera Frenkel*. Berlin: Hatje Cantz Verlag, 2013.

Tuori, Martin, and Michael Century. "As Real as What? The Stakes in Virtual Reality." Canada, Department of Communications, Conference on Culture and Technology, Montréal, 1991. Banff Centre Archives 1993-62, Box 2.

Turchet, Luca, Rob Hamilton, and Anil Çamci. "Music in Extended Realities." *IEEE Access* 9 (2021): 15810–15832.

Turkle, Sherry, and Seymour Papert. "Epistemological Pluralism and the Revaluation of the Concrete." *Journal of Mathematical Behavior* 11, no. 1 (1992): 3–33.

Turner, Jeremy. "Sat-Tel-Comp—SSTV at Open Space 1978–1982. Pioneering the Historical Paradigm for an Emerging Global Media Arts Network." Paper presented at the First International Conference on the Histories of Media Arts, Science and Technology, Banff, 2005.

Utterson, Andrew. "A Computer Animated Hand." Library of Congress, https://www.loc.gov/programs/static/national-film-preservation-board/documents/computer_hand2.pdf.

Utterson, Andrew. "Early Visions of Interactivity: The in(Put)S and out(Put)S of Real-Time Computing." *Leonardo* 46, no. 1 (2013): 67–72.

Valiquet, Patrick. "Animating the Object: Marcelle Deschênes and Acousmatic Education in Quebec." *Organised Sound* 22, no. 3 (2017): 385–393.

Vallee, Mickey. "Glenn Gould's *The Idea of North*: The Cultural Politics of Benevolent Domination." *Topia: Canadian Journal of Cultural Studies* 32 (2015): 21–41.

Waldrop, M. Mitchell. *The Dream Machine: J. C. R. Licklider and the Revolution That Made the Computer Personal*. New York: Penguin, 2001.

Wallace, Keith. "A Particular History: Artist-Run Centres in Vancouver." In *Vancouver Anthology: The Institutional Politics of Art*, edited by Stan Douglas. Vancouver: Talon Books, 1991.

Waterman, Ellen. "Improvisation and the Audibility of Difference: Safa, Canadian Multiculturalism, and the Politics of Recognition." In *Negotiated Moments: Improvisation, Sound, and Subjectivity*, edited by Gillian H. Siddall and Ellen Waterman. Durham: Duke University Press, 2016.

Waterman, Ellen. "*Patria* at the Millennium." *Topia: Canadian Journal of Cultural Studies* 6 (2001).

Weibel, Peter, ed. *Beyond Art: A Third Culture: A Comparative Study in Cultures, Art, and Science in 20th Century Austria and Hungary*. New York: Springer, 2005.

Wein, Marceli. "Twenty-Four Years of Computer Graphics at NRC." Paper presented at the Graphics Interface, Halifax, 1990.

Wein, Marceli, and Nestor Burtnyk. "Computer Animation." In *Encyclopedia of Computer Science and Technology*. New York: Marcel Dekker, 1976.

Wells, Paul. *Understanding Animation*. London: Routledge, 1998.

Wershler, Darren, Lori Emerson, and Jussi Parikka. "Lab Space: Situated Practices in Media Studies." University of Minnesota Press. https://manifold.umn.edu/projects/the-lab-book.

Westerkamp, Hildegard. "Acoustic Ecology and the Zone of Silence." *Musicworks* 31 (1985).

Westerkamp, Hildegard. "Exploring Balance and Focus in Acoustic Ecology." *Soundscape: The Journal of Acoustic Ecology* 11, no. 1 (2011).

Westerkamp, Hildegard. "Linking Soundscape Composition and Acoustic Ecology." *Organised Sound* 7, no. 1 (2002).

Westerkamp, Hildegard. "Soundwalking" (1974). In *Autumn Leaves: Sound and the Environment in Artistic Practice*, edited by Angus Carlyle. Paris: Double Entendre, 2007.

White, Kenneth. "Strangeloves: From/De la région centrale, Air Defense Radar Station Moisie, and Media Cultures of the Cold War." *Grey Room*, no. 58 (Winter 2015): 50–83.

Whitney, John. *Digital Harmony: On the Complementarity of Music and Visual Art.* Peterborough, NH: Byte Books, 1980.

Williams, D. C. "Acoustic Space" (1953–1957). *Explorations: Studies in Culture and Communication* 4 (2016).

Wisnioski, Matthew. "Why MIT Institutionalized the Avant-Garde: Negotiating Aesthetic Virtue in the Postwar Defense Institute." *Configurations* 21, no. 1 (Winter 2013).

Wisnioski, Matthew, Eric S. Hintz, and Marie Stettler Kleine, eds. *Does America Need More Innovators?* Cambridge, MA: MIT Press, 2019.

Wolf, Mark J. P. "A Brief History of Morphing." In *Meta-Morphing: Visual Transformation and the Culture of Quick-Change*, edited by Vivian Carol Sobchack. Minneapolis: University of Minnesota Press, 2000.

Wolff, Christoph. *Johann Sebastian Bach: The Learned Musician.* New York: Norton, 2000.

Young, Gayle. *Hugh Lecaine, Pioneer in Electronic Music.* Ottawa: National Museum of Science and Technology, 1989.

Youngblood, Gene. "Cinema and the Code." *Leonardo* 2, SIGGRAPH special supplement (1989).

Youngblood, Gene. *Expanded Cinema.* New York: E. P. Dutton, 1970.

Yuxweluptun, Lawrence Paul. "Inherent Rights, Vision Rights." In *Immersed in Technology: Art and Virtual Environments*, edited by Mary Anne Moser and Douglas MacLeod. Cambridge, MA: MIT Press, 1996.

Index